Mastering the Art of Beekeeping

Mastering the Art

# of Beekeeping

Volume Two

by Ormond and Harry Aebi

Illustrations by Eric Mathes

PRISM PRESS

Published in 1982 by

PRISM PRESS
Stable Court,
Chalmington,
Dorchester, Dorset DT2 0HB.

© Ormond and Harry Aebi 1979, 1982

First published in U.S.A. 1979 by
Unity Press, Santa Cruz, CA 95060.

ISBN 0 907061 23 0 Hardback
ISBN 0 907061 25 7 Paperback

Printed by Plymouth Web Offset Ltd., Plymouth.

To ALL who enjoy the great out-of-doors and especially to those who think kindly and lovingly of our indispensable, industrious little friends the honeybees, this book is joyfully dedicated.

Table Of Contents

Illustrations

Back Cover: 268 lbs. of Honey from Hive No. 4 in 1976 Harry on the right, age 86. All other factors being equal, great honey production requires much spring storage space for nectar.

Foreword

THE GRACIOUS FAVORABLE RESPONSE to our first book *The Art & Adventure of Beekeeping* has been most gratifying to my father and me. We have received many beautiful letters, telephone calls, and visits from folks living in all parts of the United States as well as fan mail and visitors from foreign countries. Many of these delightful letters and visitors have asked us for additional information on keeping bees, and this present book has been written in response to these many inquiries.

Ever increasing numbers of people are intrigued at the thought of keeping a few hives of bees. Having become conscious of the fact that in many localities honeybees exist in their own immediate neighborhoods, they wonder why they too cannot keep bees to obtain honey for their own households. They eagerly make inquiry as to the cost of bees, beehives, and other necessary equipment. Most of them quickly decide to buy the equipment, but why not hive their own bees and save that thirty-five to forty-five dollar initial expense? Usually they know of a bee tree, house with bees in the walls, or even a cluster of bees hanging on a nearby tree, but they feel a need for just a bit more instruction before trying to hive a swarm single-handed, as it were, and alone. Thus an astonishing number of prospective beekeepers write to me or telephone seeking encouragement or guidance.

Beekeeping is a tremendously large subject due to the fact that there are so many variable factors involved. Bees can live in many parts of the world from sea level to above seven thousand feet elevation, and from the tropics almost to the Arctic Circle. With a more extensive knowledge of bees and proper handling, many a beekeeper can encourage his bees to produce an even greater excess of honey per hive. This new work is an in-depth discussion of the ways of the honeybee and how to handle them for the greatest possible profit and enjoyment.

For their gracious help and encouragement we wish to express our deep gratitude to Catherine Banghart, beekeeping editor extraordinaire, and to Dr. Harold Sundean, M.D., Margaret Haun, Suzanne Gouldie, Marjorie Hillestad, Gary Blankenbiller, Greg Smart, Lee LeCuyer, Mathilde Sivertsen, Fred Stoes, Father Francis Markey, Lee Elliott of the University of Hull, England, Dr. William Abler of the Illinois Institute of Technology, Richard and Evelyn Dunstan, and Pastor Bill Creecy, Donna, and Karen. And to all other friends and beekeepers who have urged us to write this book, our heartfelt appreciation and thanks.

PUBLISHER'S NOTE: We have used the symbol of a bee 🐝 throughout the text to indicate that following this symbol are the tips and secrets offered by the Aebis for the success and joy of beekeeping.

1

Bless Those Eager Beekeepers

Hiving a Swarm at Night

On MONDAY EVENING March 14, 1977 my telephone rang.
A young female voice charged with suppressed excitement
asked, "May I speak to Ormond Aebi?"

"Ormond, speaking," I answered.

"Oh, good. We've read your bee book *The Art & Adventure of
Beekeeping*—and we have a swarm of bees. Can we hive them
in the dark?"

"No!" I looked out of the window as I spoke. The whole sky
was covered with heavy clouds in the evening darkness and a
chill wind was blowing. "Always hive bees in the daytime when
the sun is shining."

"But we have to hive them tonight," the eager voice continued. "We acquired one hive last year and we think this swarm
is from our own hive and it's hanging from a tree in our neighbor's backyard—and our neighbor is just scared to death of
bees. We absolutely must hive them tonight and move them to
our place the first thing tomorrow morning. What can we do?"

"I've never heard of bees swarming so early in the season," I told her. "Are you sure you have a swarm of honeybees out there?"

"Yes, I am," she replied, "and a really big swarm too—and we must hive them tonight—really we must. Can you tell us what to do? We've never hived a swarm before and are a little uncertain how to begin."

I caught my breath and at first thought almost groaned aloud. If one wants to get a stinging one of the best possible ways to do it is to try hiving a swarm of bees at night. At night bees crawl everywhere, we squeeze them in the dark, and they sting. But then I realized that this young woman had a delightfully severe case of "bee fever" that eager irrepressible desire to acquire more bees. I began to consider giving her instruction so that she could hive the swarm with a minimum of risk to herself or discomfort to her bees. I took another look out of the window.

"But it's too dark and cold to hive bees tonight. Better wait until tomorrow. But if you *must* hive those bees tonight, let me think a few minutes." And then a thought struck me. "You're not calling long distance, are you?"

"Yes, I am," she laughed. "We live in Walnut Creek away east of San Francisco and it's not nearly as cold here as where you live, so near to the ocean. Really, it's quite nice here this evening."

"Oh, gracious! Well, here come the directions quick as I can give them. But first, where is this swarm located on the tree?"

"About three to four feet above the ground with the cluster massed around the tree trunk at the parting of the first large limb. There's no way to shake them off onto a hive."

"Do you have a hive all assembled and ready for bees?"

"Yes. We got it all ready last year in hopes of catching a swarm but were never able to find one."

"Good! Can you also find a large piece of corrugated cardboard?"

"Yes, and everything else that we'll need if you can just tell us how to proceed."

🐝 "Take the cardboard and about halfway along one side of it cut out a notch that will fit the trunk of the tree as snugly as possible about two feet up from the ground. Block up the cardboard in as secure and as nearly a level position as possible. Place your beehive on the cardboard with its entrance facing the tree. Use small lengths of narrow board to completely block off any possible way for the bees to crawl under the hive, which they'll surely do if you're not particularly careful on this point. Then remove the cover from your beehive. Next, take a dipper and dip off a quantity of bees from the tree and dump them over the frames in your hive."

"We don't have a dipper," said the eager voice.

"Well then get a kettle of about two or three quart size that has a handle sticking out at one side and use that for a dipper."

"Oh good! We've got that. Should we use a flashlight or a floodlight?"

"A flashlight. And use that sparingly. You don't want the bees to fly. You want them to crawl. If you use a floodlight too many of the bees will take wing and fly all over the place and when you finish hiving and shut off the lights they'll be lost out in the dark and cold and will die before morning."

"Continue please. We're with you."

🐝 "Have your smoker going and from time to time smoke yourself to kill your body scent. Use your smoker sparingly on the bees, only enough to quiet them if you must, or to get them to move. You want to accomplish the whole operation quickly and quietly. Gently dip and brush off from the tree as many bees as you can and dump them on top of the frames. Empty a few smaller lots on the landing in front of the entrance to set up fanning to lure others that are crawling around on the tree trunk or cardboard to come to their new home. Replace the hive cover yet tonight if you're able, that is, if the masses of bees you've emptied onto the frames sink down sufficiently between the frames. Otherwise cover the hive as best you can to protect the bees from possible rain or frost tonight and finish

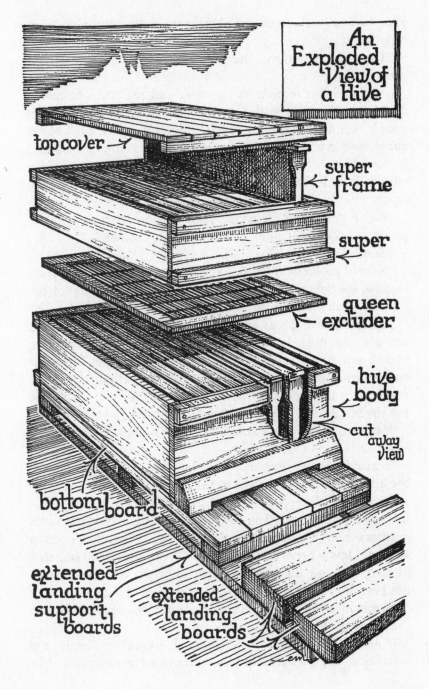

An Exploded View of a Hive

top cover

super frame

super

queen excluder

hive body

cut away view

bottom board

extended landing support boards

extended landing boards

your hiving in the morning. You're on your own. Good fortune!"

"Thank you so much!" the happy voice responded and hung up.

I replaced the receiver on my end and sat down to rest. I could not help but shake my head in dismay. What young people will not risk to get a hive of bees! But then I felt better about it—I had done the same thing in *my* younger days. I fervently hoped they would succeed. For the next hour I mentally followed them step by step. Then the telephone rang again.

"We've got them!" A young man's voice, charged with repressed excitement, almost made the wire crackle.

"Good! I'm so glad!"

"No real problem in the hiving, but we can't get the cover on because the bees are still about three inches thick on top of the frames. What shall we do?"

"Lean the hive cover up against the hive body, and over it and also the hive, lay a rug or gunny sack or most anything you have handy that will afford the bees some degree of shelter and warmth until morning. By that time they'll probably have gone down between the frames sufficiently for you to replace the cover. Did the lady who phoned me make out well too?"

"My wife is listening in on the extension phone. You can ask her."

"I made out just fine too," she laughed. "And we got the bees! We'll move them home in the morning."

Wonderful young people, I thought, the kind that really make this country so great. The next morning the young man was gracious enough to telephone me again to say that the bees had all gone down between the frames during the night and that he had replaced the cover without difficulty and had moved the bees home. I cannot help but marvel at the good-natured determination with which many a truly novice beekeeper begins the keeping of bees.

In all seasons we must be as aware of, and responsive to, the weather as our bees are. Seldom will bees swarm on other than a sunny day, but they sometimes do swarm just before a change in the weather such as a late spring shower. Then if we are called to hive a swarm hanging from the limb of an apple tree or comparable place we must use our smoker with care to smoke and warm up the bees before they will move and crawl toward and into our hive box. Such hivings are difficult but they can be done by using skill and patience.

We must remember that at all times of the year in all kinds of weather be it rainy, cold, snowy, or hailstorm that our bees must have air. Let me illustrate.

Adjusting to the Weather

The fall of the year 1975 was exceptionally dry. Only a few inches of rain fell during the months of October, November, and December. It seemed that our winter had turned into summer. The wind was almost always calm, the air balmy, the sky sunny. All the trees and plants trying to grow on the vacant lots and fields and parks suffered from a lack of moisture. Our garden area was as dry and hard as a bone. The old cornstalks that should have been rotting were all intact. Our bees flew every day but there was not much for them to gather. Nevertheless they did manage to find some pollen of various shades and hues and a very small amount of nectar, enough to convince them to allow the drones to remain in some of the hives. Never before in our lives had my father and I seen drones in any hive after November 1, but now there were a few drones in hive No. 4, our world's record-breaking production hive of 1974, and also in hives No. 5 and No. 8. With great interest we watched to see how long the bees would tolerate the presence of drones. December 1 came and went and we saw drones. Our friend Jim, who had eight hives of bees, came over and told me he also had drones in two of his hives. Again on January 1 he came to visit us and reported that he still had drones in those two hives. We watched awhile and also saw drones in our three hives. February 1 was a beautiful sunny day and again we saw drones.

February 4 was warm and a little muggy. Toward evening the

air turned colder and large clouds appeared. It surely looked like a coming storm—and we fervently prayed for one as we needed the moisture so badly—but there was no wind. Sometime after midnight we were startled by a tremendous clap of thunder which shook our old house until I wondered if it would ever settle down again. The noise even awakened my father and that is quite an accomplishment as he is a real sleepyhead as well as being quite hard of hearing. Again we saw a shivery flash of lightning that for a few brief moments weirdly lightened the entire landscape. Seconds later the thunderclap again shook our house. And then it began to hail. How long it hailed I do not know for I went back to bed and dozed off. When I awoke, startled into wakefulness by a severe cramp in my stomach, I glanced out of the window. Everything was a glorious white in the light and reflection of a street lamp. Snow—in Santa Cruz? I jumped into my clothes, donned a light coat and hat, and ran outdoors to look at our beehives. It was still hailing lightly, snowing a little, and almost raining, all mixed together. I looked at our nearest beehives. They were blocked with hail and snow and the entrances were all sealed shut. Feverishly, with freezing cold hands, I scraped the snow and hail from the first entrance so that the bees could get air. The draft of warm air from the bees inside the hives had caused the hail to solidify into a hard mass and it took longer to completely clear an entrance than I had anticipated.

I soon gave up trying to entirely clear one hive before I went on to the next. After I had poked an air hole through the entrance of each hive I went back and scraped each landing board clean. By the time I had finished it was six-thirty in the morning and I went into the house wet and cold, but successful, for all of our bees survived.

Some of the hail remained on the ground all day Thursday. My father and I went down to the shore of Monterey Bay, only three blocks away, to see if there was snow on the sand. Indeed there was, and that was a sight never before seen in this country as far back as anyone could remember. We went home and wrapped our beehives in sheet plastic for we knew the coming night would be very cold, at least for this country, and we did not want our bees to lose any of their brood, as the weather had been so balmy that the queens of every hive had laid many eggs and the bees were covering much brood. As it turned out all went well with every hive. But we will always remember the great hailstorm of February 5, 1976.

Hiving High Swarms

My friend Lee LeCuyer bought his first hive from me in the spring of 1975. We had several lengthy chats on the subject of beekeeping and then he bought a copy of my bee book and read it through. He had no more than finished reading it before he told me that he wanted more bees and that if I heard of a swarm hanging somewhere that I did not want for myself he would be glad to go and hive it. Well, I did not have to find him a swarm as within a few days an acquaintance of his telephoned him to come get a big swarm. So Lee drove out into the country with all of his hiving equipment loaded into his station wagon and was prepared, as he supposed, to hive any swarm available. When he got out to the place he found a truly huge swarm but it was hanging from the limb of a tree fully fifteen feet above the ground—and he had only a six foot step-ladder. He should have come to see me and secured a good ten or twelve foot stepladder and another ladder to lean against the tree for him to stand on while he worked. But Lee figured that he did not have that much time for the swarm had already been there for some hours before he had heard of it and it might take a notion to fly away across the valley or over the hill without asking his permission. So he put his mind to the problem, found an old rickety ladder to climb up on, and then placed his too short stepladder under the swarm. On top of this he placed an old box, and then another on top of that, and on top of it all he carefully and gingerly eased up his hive, balancing it nicely on the uppermost box. A gusty wind shook the tree from time to time agitating the swarm of bees. Lee had his hands full keeping an eye on the swarm, trying to balance on his own ladder, and at the same time trying to keep his smoker going with his free hand, and also balance the teetery beehive. All seemed to be going well until just an instant before he planned to shake off some of the bees onto the top of his hive. At that moment a strong gust of wind caught and shook the tree. His rickety ladder began to give way and to save himself he had to drop his smoker and release his hold on his precariously balanced beehive. Down went the smoker followed by his carefully prepared beehive—and the heavy beehive landed—wouldn't you know it—right on top of his smoker almost smashing it flat—and completely cutting off its snout.

Lee surveyed the wreckage. His hive was not too badly damaged but his smoker was a wreck. However, he straightened it out as best he could and found that it would still smoke. There was no way to direct the smoke but at least it would smoke enough to be of *some* use. So he decided to go home, bring back a strong ladder and other things he would need, and try again. This time he had better success and was working happily to get the huge swarm all hived—when half of the mass of bees took wing and flew off across the valley—and out of sight. Lee was disappointed, of course, but he succeeded in hiving the remaining bees and took them home. Elapsed hiving time, seven hours, and almost forty miles of driving. The next day he came to see me.

"Ormond!" he exclaimed, "you'll never believe what I've been through!" And then he went on to explain all that had happened.

"What do you think of that?" he asked me.

"I think you caught half of a double swarm—and you should be glad. And I think you should rejoice that you didn't fall off that old ladder when the wind struck you, for you could have broken your neck—and necks are too short to splice!"

"You're right, and I *am* glad," he nodded soberly. "And I did get my first swarm—such as it is," he added with a note of justifiable pride in his voice.

"Right!" I agreed.

Lowering a High Swarm

As time passed Lee became an efficient beekeeper harvesting almost 100 pounds of honey per colony from more than a dozen hives. About the middle of May 1977 Lee telephoned me.

"One of my hives just swarmed," he said. "And would you know, those little beggars passed over every low bush and tree in my big yard and clustered away up in the tiptop of my neighbor's big walnut tree. Do you think they'll come down?"

"Not likely," I answered.

"Well then do you think that if I climbed up the tree and cut off the small limb they're hanging from and let it drop to the ground, that they'll recluster somewhere else, hopefully down lower where I can get to them more easily?"

"No, I don't think they would, not in my experience. They'd just take wing and go flying up to the original area again. But let me think a moment," I said. "Do you have a hive ready to put them in if you could get them down?"

"No, I don't have a hive or anything else ready," he answered rather dejectedly. "Do you have one you could sell me right quick?"

"Not really," I told him, "for I'll need it for myself one of these days. But wait a minute. I know a man who's most anxious to acquire a swarm. Would you sell that swarm if you could get it down?"

"Yes, I'd sell—for eighteen or twenty dollars."

"Let me phone the man. I'll call you right back."

I put in a call to Lloyd Coleman of Ben Lomond, a small town about seven miles away.

"Would you want to pay eighteen dollars for a swarm of bees?" I asked. "The owner and I would help you hive them and it's worth about that much to see how it's done—and you could never hive these bees yourself without extensive experience. Or had you hoped to get a free swarm?"

"I'd hoped to find a free swarm. But I've never hived a swarm of bees and some instruction would help. Let me talk it over with my wife, and I'll call you right back."

"We'll take them," he reported minutes later.

"Then come down to Santa Cruz as fast as you can—but drive carefully—and I'll meet you at the owner's place." And I gave him the address. "I'm going to give you boys a little help."

Then I telephoned Lee. "The man will give eighteen dollars for your swarm and I'm coming right over to help you hive them."

"Good!" Lee exclaimed. "The walnut tree is growing on the side of a ravine and it's brushy down there under the tree. I'll try to find the best place to set the hive."

Time is of the utmost importance when going for a swarm of bees. A cluster may stay in its original location for only fifteen or twenty minutes—or as long as two weeks—more often the former. So I loaded my station wagon as fast as I could with a thirty foot extension ladder and also an eight foot stepladder, my smoker, veil, bee brush, some rope, and a handsaw. In twenty minutes I was at Lee's place.

"They're up in the top of that big tree," he pointed toward the top of a great leafy green crowned English walnut tree. I looked but could see no bees.

"They're up there," he said. "Let's go around through my neighbor's lot and get under the tree. You'll be able to see them from there."

And then I did see them—and they were high—higher than I could reach with my extension ladder. The large cluster was hanging amid the leaves and twigs from a one inch diameter branch that bent threateningly under the weight of bees.

"If that branch would only break," Lee said, "those bees would come down of their own accord."

But green and growing walnut branches are very strong and seldom if ever break. So we had no hope of getting the bees down of their own accord. Lee was unhappy with his bees and as he talked I had an opportunity to really study the situation. Walnut trees are self-pruning in that all small limbs shaded by the great green canopy overhead soon die, become brittle in a year or two and may be brushed aside as one climbs the tree.

Only the main trunk branching limbs were alive and they were of sufficient number so that with the aid of my stepladder to reach the first low limbs I could climb the tree.

"I'm going to climb that tree and lower those bees down to you," I told Lee.

He looked up at the bees. "Be careful!" he admonished. "Those dead limbs break without any warning, and those bees are away out over nothing. That ravine is deep and there is nothing between them and the bottom!"

I was aware of that too, but the tree could be climbed, so we went to my car for the stepladder. There we met Lloyd Coleman who had just arrived. Lloyd and Lee recognized each other. They had met on a bee adventure the year before. We carried our equipment including Lloyd's beehive and my handsaw and two lengths of rope down to the foot of the tree and I began the ascent. I wore my bee veil as I climbed as I had no free hand to carry it. I needed no other protective gear. It was a great help that Lee and Lloyd looked up from below and could direct me as to which of the longer branches were dead so that I could beware of them as I climbed. I reached the top of the tree safely and found that I could just barely reach the small branch upon which the bees were clustered. It was bending dangerously so I hurriedly tied myself to the thin main top trunk of the tree leaving myself on a six foot tether, as it were, so that I could reach out over the void toward the bees without danger of dropping forty feet or so if I slipped and fell. Not that I relished the idea of falling and being caught by a rope around my chest and under the armpits for that would be a most painful way to be stopped short in a free fall, but it would be better than to risk a sudden plunge wherein I could break my neck.

After some maneuvering I managed to tie a quarter inch nylon rope around the branch holding the cluster and then with the handsaw in my left hand and holding on for dear life with my right hand I sawed the limb off. The sudden short drop as the limb and bees started to fall before the slack was taken up by the rope around my left wrist caused the cluster to disintegrate into thousands of bees falling toward the ground. Some of them actually did fall onto the ground but most of them took wing before they had fallen more than twenty feet and came up again with a great swoosh all around me in an effort to recluster. Hundreds of bees alighted upon me for a moment but there was no danger as such bees never sting, at least not in my experience, nor did these. I raised and held the cut branch in my left hand into as nearly the same position as it had been while on the tree and the bees immediately began to reassemble again around the quart-sized mass that had remained on it. I hoped that the queen had not been dislodged and from the way the bees reclustered I was sure that I still had her on the branch I was holding. In three or four minutes the branch became very heavy and I called to the men below that I was going to lower it down. They were ready, caught the branch with its cluster of bees, and shook them off over and in front of the open hive. The bees began to go right in. Then I drew the branch up to me again and held it for a few minutes so that the circling bees could again cluster upon it. This they did without loss of time, attracted by the lingering odor of the queen's scent. Once more I lowered the branch and the men below shook them off over the hive. A third time I drew up the branch, the last of the bees soon clustered upon it, and down they went. Lee replaced the cover on the hive and called up to me that we had them. I climbed down safely and Lloyd had his much desired hive of bees.

The Affluent Approach to Hiving High Swarms

If a hive swarms and the cluster settles thirty or forty feet up in the top of a tree too straight and slender to climb, we may secure the swarm in several ways. Provided it is possible to get a wheeled vehicle close under the tree we can do as a friend of mine did: he hired a telephone lineman's vehicle with a power bucket hydraulic lift arm that lifted him and his beehive up within easy reach of the swarm. In half an hour he had the swarm hived and was lowered to the ground. Then he had himself lifted up again so that he could place a catcher hive in the exact position where the swarm had been. He hung the catcher hive in place with a rope passed over the limb, and after being lowered to the ground, he tied the other end of the rope around the base of the tree. He used a rope long enough so that when another swarm took possession of the catcher hive he could lower it to the ground. Renting such a vehicle was expensive but he said it was worth the money because within a week he had caught another beautiful big swarm and he could lower that one down by himself. He was tremendously pleased with his two new swarms.

Bees have their flyways much the same as ducks and geese, but on a much smaller scale. Where one swarm of bees has clustered we may later in the season find another swarm, for the queen bee leaves a scent that lingers for days or even weeks, depending upon the weather. So if we have been away from home on a warm day during the swarming season, upon returning I always glance first at the places where bees have clustered in times past and then check elsewhere in our yard. On occasion we have come home to find a wild swarm clustered on a previously scented bush or tree limb.

The Radical Approach

Another method of securing a high swarm as recommended to me by our former bee inspector, now retired, is to stand almost directly under the swarm in the tree and with a shotgun shoot the limb off between the cluster and tree trunk, dropping limb, bees, and all, and then hiving the bees into one's waiting hive box.

"You'd kill the queen—dropping her from a height of thirty or forty feet!" I exclaimed.

"Never!" he said. "I've done it about fifty times and never killed a queen yet. The queen always survives though one does kill a few dozen workers who get crushed under the limb and heap."

"I don't see how she could possibly survive," I insisted.

"She's always in the center of the cluster and the many bodies of the worker bees always cushion her fall. I've had no trouble hiving such a fallen swarm."

I had to believe him for that old boy knew a vast lot about bees. But I have never had the opportunity to try his shotgun method.

Luring and Lowering a High Swarm

Recently Paul and Dorothy Ames of Ames Apiaries of Arroyo Grande, California, explained to me their novel method of bringing down high swarms if there is a beehive located nearby. They wrap a considerable length of light rope into a double-fist-sized ball and throw the ball of rope up and over the limb upon which the bees are clustered, retaining the free end of the rope in one hand. After the ball of rope has passed over the limb it unwinds as it falls back to the ground. Then the Ames quickly remove a frame of brood from a nearby hive, brush the bees from it, tie it to the rope and pull it up to the immediate vicinity of the clustered swarm. In a few minutes the bees all crawl onto the frame of brood and the Ames lower the swarm and hive it. They say that every last bee will come down on the frame of brood. This method I would love to try!

Tending the Honey Flow

There is a tendency to become so engrossed in hiving early swarms that we forget to service a hive already acquired. A large new swarm may draw the ten brood frames in from five to fifteen days so we must remember to observe the bees occasionally and give them a queen excluder and super when needed. We do not want our new bees to quickly become overcrowded and, in their turn, swarm out. We want them to build up strongly for the purpose of producing excess honey for us as beekeepers.

If during the late winter or spring honey flow we find that we must be away from home and can neither add nor remove supers as needed to our strong new or established hives, we may still secure a goodly amount of honey. We remove the en- trance closure cleats until the hive entrances are about three- quarters open and on the warmest day available place three medium depth supers on each hive above the queen excluder. We place a few heavy rocks, at least forty pounds, on top of the roof cover of each hive and go on our way rejoicing. Upon re- turning home we often find that the bees have been busily at work in our absence and have the hives jam-packed with honey. There is some risk in this procedure and one does lose a hive now and then, usually due to robbing, but my father and I, as well as others, have found that the rewards are usually far greater than the risks. We do try to make arrangements with a nearby beekeeper to have him observe our bees from time to time during our absence to see that no vandals, skunks, or ants are molesting the hives. But other than that we let the bees take care of themselves until we return.

I always remember the man who had little success with his bees until he left for a six month's trip to Europe one spring and summer. He placed all of his available supers on his three hives and departed. When he returned he was astonished to find all of his hives full to overflowing with honey. The bees had built cells and crammed honey into every available space between the frames and everywhere else. It was difficult to remove the beautifully filled frames, but at least he had honey—more honey then he had seen in all the years previous put together.

"You know," he exclaimed to me, "I should take a trip to Eu- rope more often!" I laughed and agreed with him.

Hiving Established Wild Bees

In days long past when I myself was still more or less of a be- ginning beekeeper I received a telephone call one noontime to go away out into the country to hive a swarm of bees.

"They're in an old chicken house," my caller informed me.

"Is it a swarm?" I asked. "And if so is it hanging from a rafter, or a partition, or is it clustered on a wall?"

"None of those places," he said, "but in a cranny near the

end of one wall almost at floor level. I don't know if it's a new swarm or not as I just rented the building and found the bees there. Can you come and get them?"

"Yes," I replied, "my father and I want a few more swarms at this time." So we quickly loaded our station wagon with everything we thought we would need and started out. It seemed strange to us that there would be bees in a chicken house for we could think of no good place for them to build their nest, and a new swarm recently emerged from a beehive had never, in our experience, ever clustered in a chicken house. In due time we arrived at the ranch and the man showed us into the chicken house and over to where the bees were busily coming and going flying in and out through a broken windowpane. Then I saw where the bees had made their home—in an abandoned hen's nest, second tier of nests up from the floor. In dismay I knelt down and looked into the nest box. I saw immediately that the bees had been there for some months at least for the nest cavity was almost completely full of drawn honeycomb to the point where the bees were becoming crowded. Every comb was covered with bees busily at work. I stepped back to discuss this unusual situation with my father.

"Those bees are going to be exceptionally tricky to hive!" I whispered in an urgent undertone to my father so that the man who had gone toward one end of the chicken house could not hear us. "What shall we do?"

"If it's left up to me," he answered, "those bees will stay right where they are until doomsday. I'm not sticking my hands into that hot spot!"

"But we've come a long way up here into the mountains to get these bees," I told him, "and they seem quite tame—I'd like to give them a try."

"Do as you wish," he said, "but you'll have to cut out the combs and tie them into the brood frames just as you would a free-hanging hive. I'll help with the tying but that's about all."

In a few minutes we had our hive set up on the floor about fifteen feet away from the hen's nest so that we would be more or less unhindered by bees while we were tying lengths of cotton string around the pieces of honeycomb I planned to cut and fit into the frames. I started up my smoker and carefully brushed the bees from the outermost comb into a long-handled kettle that I found nearby. We had not come equipped to hive such a swarm. I set the kettle with its bees close in front of the hive entrance. Then I cut off the first honeycomb, caught it in both hands, and carried it to our hive where after brushing off the bees from both sides over and onto the five frames still remaining in the brood box, we fitted and tied the honeycomb into an empty frame. This done we placed the frame in our brood box and I dumped the bees from the kettle into the box over the honeycomb. The poor bees appreciated even this semblance of a home. But by the time I had cut loose and removed the second bee-covered comb the bees in the nest box were really stirred up.

"Look in our bee equipment box," I called to my father, "and toss me the household rubber gloves you'll find there." My father looked as requested.

"I don't find any rubber gloves in here—or any other gloves, for that matter. You must have left them at home."

"Look again!" I urged. "These combs go away back out of sight in this dark nest. This hen's nest must be oversized or something, and these bees will sting because it's so far back and dark in there that I can't see what I'm doing."

"No gloves." My father stated the grim fact with finality. "You're on your own. And you really have those bees stirred up. Now you *have* to hive them—there's no stopping now." What he said was only too true.

"Oh Lord," I prayed in a hoarse whisper, "shut the lion's mouths!"

And He did! God answers prayer! Again and again I reached bare-handed as far as I could into that nest hole and one by one cut off and removed the bee-laden honeycombs, pulling them out of a veritable inferno of flying, crawling, and buzzing bees. And though I many times felt the pressure of my hands and

Those
Beloved
Honeybees

fingers pressing momentarily upon many bees, yet I was able to shift fingers quickly enough to relax the pressure upon any one bee until it could make its escape and I felt my fingers press the honeycomb itself, at which point I could cut it off and remove it. My veil got shifted around awkwardly so that from time to time another bee crawled up my neck and into my veil, but they too did not sting me. Often and often I felt a bee crawling up my pant legs and I had to keep moving and stamping my feet to relax the pressure of my clothes against a bee and cause it to crawl back down and out. At the height of the activity the man who had called me came in to look—from a discreet distance—of course.

"Say," he said, "that's the first time I've ever seen a beekeeper dance the Charleston! You're really pretty good! Getting badly stung, I'll bet?"

I shook my head in the negative—I was too busy to talk. In due time I had all of the combs removed and my father and I finished our hiving. We loaded our precious bees and started for home, leaving behind a truly astonished householder—and he was no more astonished than I at the successful outcome. And I had not been stung though the floor was covered with bees and I had often had to kneel among them, reach far into the nest box among them, and literally had them flying and crawling all over me for the better part of an hour.

Tips on Moving Hives a Short Distance

An Evangelist friend of mine, Reverend Vernon McGrew of the Free Holiness Church who lives near Marble City, Oklahoma, telephoned me long distance on May 25, 1977 with two problems in beekeeping. He had bought a copy of my first book *The Art and Adventure of Beekeeping* the year before when he was passing through Stockton, California. Being much pleased with the book he telephoned and asked if he could come to see my father and me. He and his family were on their way to Fresno and would not mind detouring the few dozen extra miles to see us before they reached their destination. I told him we would be glad to see them. We had a lovely visit and he expressed a great desire to get started with bees. We discussed the possibility of getting bees out of a tree in his home area, the wall of a house, or buying package bees. Now his telephone call from Marble City clued me in that he had somewhere found some bees.

He told me that a few weeks earlier he had let it be known among his friends and acquaintances that he wanted bees. Two days prior to his telephone call, while he was away from home trying to remove and hive a swarm from inside the wall of a house, a friend of his not finding him at home, had set off in front of his house a box full of bees and left them there for Vernon to find when he returned home that night. The friend had dropped them off in the forenoon but Vernon had no immediate way to know when the box had been left at his house. He did not want the bees permanently located where he found them so after dark he picked up the box and moved it several hundred feet toward the back of his property. The next day, to his surprise, he found that by noon a great swirling accumulation of bees was flying around and above the spot where the hive had originally been set down. Vernon realized that these bees were fielders who had been out foraging the day before. 🐝 At sunup the bees had flown out of the hive thinking they had already oriented themselves as to their hive location but they had not taken into consideration that Vernon had moved them during the night for he had not thought to make a low pile of grass or leaves along in front of the entrance to alert the bees to the fact that they had been moved. Such grass or leaves do to some extent restrict the entrance temporarily but as more bees crawl up and out they soon begin to pull away bits of grass or leaves and so clear the entrance themselves and at the same time become aware of their new location. But not having been warned in any way that they had been moved, they returned to the original drop-off spot and were now lost.

It was at this point that Vernon telephoned me making inquiry as to how he could get those lost bees to accept their new location a few hundred feet away. 🐝 I told him that I did not know of any certain way but suggested that he find a cardboard carton somewhat larger than a shoe box and in one end of it cut an opening about six inches long and one-half inch high and place this box on the ground where the lost bees were circling and alighting. Some of them, at least, would accept this box as a temporary shelter and he could carry them over to where his hive was now located and set it down with the entrance to the box and the entrance to the hive close to each other. When the bees crawled out of the carton some of them would enter their old home and reorient themselves to their new location. I told him he would have to let the bees crawl out, and then return his carton hive to the old location several times, and for probably two days, in order to help the maximum number of bees find where he had moved their hive.

Vernon thanked me and said he would try my idea. He later reported that my suggestion to make a "cardboard carton" hive to carry the lost fielders to his hive's new location worked well for him.

Bees in a Wall

My friend Vernon's second problem had to do with the bees he had tried to remove from the wall of the house. The owner had insisted that he open the wall from the inside because it would be so much easier to repair the wall by replacing a panel of sheetrock on the inside than to patch together the wooden siding on the outside. The owner was right about repairing the wall but he made it extremely difficult for Vernon to remove and hive the bees—as Vernon ruefully related to me on the telephone. The bees swarmed out of the wall and into the room and clustered around all the windows as soon as he had removed the panel of sheetrock. And then he and his helper got about as smoked up as a smoked herring when they tried to smoke the bees out of the wall and room—and it was an awful mess! I quite agreed with him—I have tried it. 🐝 My father and I always open a wall from the outside—and if the owner will not agree to that we do not attempt the job. After one has tried to hive the bees from inside a building and has them all stirred up it is almost impossible to then take the hive outside and hive them out there. The bees are demoralized and scattered in every direction and usually cluster in several places making it impossible to tell where the queen is and very difficult to find her. And without their queen, bees will seldom stay in the hive no matter how many times we brush or dump them into it.

An interesting letter from Vernon a few days later stated that in spite of all his problems in trying to hive that swarm from the wall of the house both from the inside and the outside, he did succeed. But within a few days he made the sad discovery that his new hive was queenless. In the three or four hour struggle to get the bees out of the wall and into his hive he had somewhere lost the queen, or injured her, or she might have flown away just hours before he began his hiving operation, if the colony had been on the verge of swarming. Sometimes bees do not wait for a young queen to emerge before the old queen and a swarm swirl out of the hive to start a new colony elsewhere. To save as many of the adult bees as possible he combined his newly hived swarm with the hive his friend had given him.

When the McGrew family had come to see us a year earlier I expressed surprise that they could find our home at night.

"No problem," he had said. "Oh, I got really lost over in Gilroy—didn't even know if I was going north or south—but I have a citizen's band radio and I asked the truckers if anyone knew where Santa Cruz was—and sure enough someone answered that he knew—and where did I want to go in Santa Cruz? I said to Ormond Aebi's, the beekeeper on Seventeenth Avenue, and he directed me right to your place. Those truckers are wonderful. There's always one of them within a hundred miles who knows all the answers."

Let us rejoice and be thankful for CB radios and helpful truckers! Vernon does what we all need to do—he uses his head when he has a problem. In the old days when my father and I had a problem my mother used to say to us, "God gave you men good heads—use them!"

Those Impossible Situations

One midsummer day Allen, who works for a large construction

company, came to see me.

"Ormond," he said, "I've found a lot of bees coming and going out of the end of an old redwood drainage culvert over on J Street. Must be a really big swarm in there. How can I get them out?"

"I don't know," I told him. "Down in the ground like that there is no telling how far they would have to go before finding an enlarged open space suitable for building their brood nest."

"Well, I can use the company's backhoe and anything else I'll need and I'm going to dig them out this weekend. I'd surely like another hive, and I just might get a lot of good honey too. They can't be very far from the end of that drain—and this is one swarm that can't fly away while I'm getting them!"

"Try it if you want to, and let me know how you make out."

Allen left with the glow of certain success lighting his face. A few days later he stopped to see me.

"Did you get the bees?" I asked.

"No," he replied in a most wistful tone of voice. "I started digging last Saturday morning and dug all day. I had to dig very carefully and slowly so as not to wreck the combs and honey when I got to them—but I never found them. I had bees flying all over the place and I dug up more than forty feet of shrubbery and parking lot and finally had to quit when I came to a traveled roadway. And it took me all the next day to refill the ditch I'd dug and repave the parking lot. But *where* could those bees have had their nest? Some were still coming out of the end of the culvert where I stopped digging."

"I haven't the least idea either," I answered.

"Well," he said, "I surely put in a hard weekend's work for nothing."

"Not for nothing," I told him, "but I think we'll have to revise the old axiom 'Don't count your chickens until they're hatched,' to 'Don't count your bees until they're hived.' "

"So true," he nodded as he turned and walked wearily away. But then his face brightened as he turned toward me and said, "But next time I might get them!"

2

The Power Of Love & Knowledge

A Lesson in Honeybee Appreciation

"I LOVE HONEYBEES and those bees are all my friends." I spoke with quiet intensity from a lifetime of experience.

"And I hate bees," a husky young man in his late teens retorted as he and two companions gathered closer around me. A fourth remained seated at a little distance.

"I hate bees, I said," the young man repeated as he glared down at me from this six-foot height. We were standing within fifty feet of an old pepper tree, hollow at the stump, which was the home of a large swarm of wild bees. As a few bees flew around in circles near us I looked steadily up at the young man who had spoken.

"Do you hate beefsteak too?" I asked.

"Of course not!" he replied, "But what have bees got to do with steak?"

"Everything. But just now I want you fellows to put out that burning torch you've lighted and leave those bees in that tree alone. They are my friends, and I don't go for this burning business."

"Well, we do, and we're going to burn out those bees—see?" He and his friends took another threatening step toward me.

"I told you those bees are my friends," I repeated as I took a step backward in the direction of my agitated little friends. "By the way, do you want to make something of this?" I asked as I noticed the man clench his fists in an attitude of challenge.

"Yeah!"

"Fine. I'll accommodate you in a few minutes." And I turned and walked into the circle of excited bees that were already in a state of agitation due to the stones and pieces of old lumber that the young men had been throwing at them.

"Are you crazy?" they called out to me as I approached the bee tree amid a veritable hum of bees. "They'll sting you to death!"

"No they won't," I replied. "I told you these bees are my friends—and they are."

The poor persecuted bees were indeed glad to see me come. They flew all around my head and shoulders and some landed on my arms and hands, at the same time emitting their joyous hum of welcome. But from time to time a dozen flew away to angrily buzz the young men, effectively keeping them at a distance of fifty feet from their tree. As I talked to my precious bees to calm and quiet them I noticed the exquisite pattern of light and shadow cast by the pepper leaves upon the bare ground under the old tree. It was a beautiful sunny afternoon in late spring.

A few minutes earlier as my father and I had parked in the general area of what we called "our bee tree" we had seen these four young men molesting the bees.

"I'm going over there and talk to those fellows," I told my father. "They need some instruction regarding bees and their usefulness to mankind."

"All right," he answered, "but be careful. Those fellows look rough and tough."

"No danger. With 40,000 little hot-tailed friends at my command we'll be able to handle the situation very nicely." And so

I now found myself safely within the circle of my beloved bees—but I still had to convince four young toughs that they should desist from their avowed intention to burn out my friends. As the bees became more quiet and reassured I walked up close to their entrance hole and placed my right hand and arm near to the opening. Immediately weary heavily laden homecoming bees began to alight on my outstretched hand and arm. Soon several dozen had stopped to rest. Others alighted more or less all over me. When I had acquired what I thought was a sufficient number I slowly left the bee tree and walked out to the perimeter of the bees' guard flight where three of the young men were lined up watching.

"Ain't you been stung?" the big guy asked in obvious disbelief.

"No. I told you these bees are my friends. By the way, did you say you wanted to make something of this?" I took a step closer toward the big fellow who had threatened me and held out my arms loaded with resting bees.

"No!! No!" he exclaimed as he jumped back a step.

"All right. Then let's talk about bees for a few minutes. Bees are our friends. They and their numerous relatives in the insect world are vital to the world's food supply. Without them we would soon all go hungrier than we are now."

"I never eat honey," one fellow remarked. "It would never affect me."

"Oh yes, it would," I answered, "that's where the beefsteak part comes in. Bees are indispensable for the pollination of the major legume crops of the world such as alfalfa, sweet clover, white clover, and many others. These legumes are the basic feed for cattle and sheep, as well as an important feed component for chickens, rabbits, and almost all other animals and fowls that we use for food."

"I thought cows ate grass," the big fellow grumbled.

"They do, but if you want steak—tender, juicy, mouth-waterin' delicious steak—you better help the honeybees to pollinate the legume crops."

"Hm-m-m," muttered the big man as he pursed his lips.

"And furthermore," I continued, "legume crops are nitrogen-fixing plants and nitrogen is a powerful natural fertilizer needed by all other crops. After a field has been in alfalfa for a few years and is then plowed up, the next crop of wheat, corn, or barley will also produce a bountiful yield. So if you want bread, pie, and cake on your tables, and almost everything else that's good to eat, help the honeybees!"

I saw two of the fellows nod their heads in approval.

"And then there's another thing," I said. "If you cruelly hurt those bees you'll most certainly suffer acute pain yourself in the days ahead. It's a law of the Universe and it never fails. No one told me when I was your age, but I'm telling you—and you'd better take heed!"

The big young man began a sneering remark when the fourth, a sallow-faced young fellow who had been sitting on the ground, suddenly got to his feet and came over to us.

"That's the truth!" he stated grimly. "Thank you for telling us. I've raised hell all my life and have caused a great deal of suffering both at home and with everyone else I've been around. Three times my father almost beat me to death trying to knock the devil out of me, but he couldn't do it. But these past two years have been hell." He paused.

"What happened?" I asked.

"First I had a bad fall and broke a leg, and before that got completely well I broke a foot, and ever since it's been one miserable unpredictable accident after the other—and all of them painful! How I've suffered!"

"What are you going to do now?" I asked as he turned to leave.

"I'm going to put out that burning torch. I was the one who was going to burn the bees—but I'll never do it again—never!" And he picked up his torch, put it out, and walked away.

"Well," said the big fellow who had first addressed me, "it's his torch and he's gone, so I guess we might as well go too."

As they walked away I heard one of them mutter, "Sure a

strange guy, walking right in among those bees."

"He should'a been stung to death. He must know some-thing—you can't be *that* lucky," another responded in reply.

"Been nice talking to you," I called after them. They waved a puzzled farewell. Then I again entered the circle of my beloved bees and went up to their nest hole in the tree where I cleared away all the pieces of broken boards and rubbish that had been thrown at them. Some of the dirt from one of the boards had broken loose and lodged in the bee tree entrance almost block-ing it. I quickly cleared the entry hole but some of the weary homecoming bees had been locked out so long that they re-mained clinging to the bark some eight or ten inches around the entrance for some time after I had it cleared. On the way home I recounted the adventure to my father.

"You know," I told him, "those fellows throwing all those boards and dirt at our bees in the tree have given me an idea. What do you think would happen if we almost entirely closed our bee tree entrance hole for half an hour and caused the field bees to again cluster in a large mass around the entry as I found them today?"

"Why would you want to do that?" he asked.

"Because we have a very weak hive at home and it seems to me that if we brought that hive over here some warm afternoon and placed it upon a platform built of those old boards, the tired field bees, not being able to reenter their tree house, might decide to crawl into the entrance of our weak beehive if we can manage to place our hive opening within a few inches of the bee tree opening. We might even place a screen cone over the hole in the tree and so prevent the returning field bees from entering without suffocating those remaining in the tree."

"That's an idea," my father agreed. "But do you think the hive queen and guards would accept the bee tree fielders?"

"That's exactly what I'd like to know," I replied. "If they would, our weak hive could be built up into a fairly strong colony in less than one hour, and the bee tree swarm is so large that they'd hardly know they'd lost a few thousand workers."

"You're right about the last part," he agreed, "and the whole scheme is a brilliant idea—if it works. Let's try it tomorrow! That whole large acreage is being bulldozed for a housing development and in a few more days they'll reach our bee tree and uproot and burn it along with everything else."

"So true," I agreed sadly, "and we don't have the time nor strength to cut the tree and save all of the bees. But with good fortune we can save some of them, and try my idea too."

Adding Wild Bees to a Weak Hive

The next day dawned sunny and warm as sometimes happens in the month of May in our Santa Cruz area if the fog does not roll in first. About one o'clock in the afternoon we loaded our understaffed beehive into our station wagon and started for our bee tree. What about the fielders returning to their hive and finding it gone? We had thought of that too and had placed a small cardboard box with an entrance hole cut into it so that the few fielders returning would have a place of shelter until we returned in approximately two hours. There is a certain amount of risk in hauling away a beehive in the warmest part of the day and leaving the fielders distraught and homeless, but in this case we hoped that the rewards would be greater than the risks. On arriving at the bee tree we found a tremendous buzzing of activity. The colony was even stronger than we had anticipated, and now on a warm day they were joyously humming with innumerable bees zooming away or coming in for a landing.

"Father," I said, "this old pepper tree has grown in exactly the right way for us to quickly build a platform so as to place our beehive with its entrance very near to the entry hole."

"You're right," he answered after he had looked at the tree. "After placing our hive all we'll need to do is fit a short piece of cardboard between the tree trunk and our hive entrance, and then install our escape cone in the bee tree hole with the screen you've brought."

In a few minutes we had everything in readiness. As the

screen and cone hindered the bees from leaving their tree, some of them tried desperately to find other exits, and for a few minutes they had their inside entry clogged with their bodies. For a few anxious moments we wondered if they would so plug their entrance as to suffocate the queen and brood in the tree, but this they did not seem to do. Meanwhile, in what had seemed to us only a few minutes, hundred of field bees returning loaded had settled all around the entry in an ever increasing mass of bees that seemed to grow by the minute. We watched with intense interest as the mass of bees approached nearer and nearer on the cardboard toward the entrance to our hive. In fifteen minutes they completely covered the intervening space. Now was the crucial test. Would they, in their dilemma at not being able to enter their own nest, consider entering our newly placed hive? Then several of them caught a whiff of the presence of bees and a queen in our beehive. They turned and went in until they were out of our sight, and others followed. We were highly pleased. But now came the second crucial test. Would the queen and her guards accept them? Within thirty seconds we saw several guards come running out of the entrance to meet the strange incoming fielders. They greeted as many as they could and seemed to approve and encourage one and all to make our hive their home. We both breathed a great sigh of relief as we watched an ever increasing march of bees follow their leaders into our hive. It was a beautiful sight.

In an hour all of the bees that had alighted around the entry to the bee tree had accepted our hive as home. So had most of those returning loaded as the time had passed. At the end of the first hour we estimated that we had acquired between five and six thousand bees, and our own bees as well as those from the bee tree were beginning to come out of our hive in ever increasing numbers and fly away again as fielders. We realized that we had acquired all of the bees we could for that day so placed an entrance closure on our hive, loaded them up, removed the screen cone from the bee tree, and went home.

There we found about one hundred field bees in and around our cardboard box which we now removed and replaced with our hive and then carefully induced the poor lost and forlorn fielders to leave their cardboard box and enter their own home. This they were most grateful to do and to our joy the newly acquired bees accepted them without question. A few days later we again visited the bee tree and found that they too were working as busily as before though their destruction was very near at hand. For us, it was a most successful experiment as it greatly strengthened our weak hive.

Tips on Wintering a Weak Hive

If our experiment had not been so successful we would have had the problem of wintering this weak hive as did our friend Lee LeCuyer a year ago. On February 28 he came over to see me bringing a beehive with him.

"Take a look at this hive, will you?" he asked. "For several weeks I haven't seen any bees come or go from it even on relatively warm days so this morning I removed the cover and looked in. All the bees are dead. What do you think caused them to die?"

As he spoke he pulled the hive out onto the tailgate of his station wagon and removed the cover, then gently lifted up a frame from the brood chamber. Clinging to the beautiful yellow wax near the top of the fully drawn comb was the queen and near her fifteen or twenty worker bees, the queen's escort. They clung so realistically to the wax that I thought for a few moments that some of them must surely be yet alive. But they were all dead as Lee had stated. Some dozens of other dead workers were clinging here and there on other portions of the combs. A careful examination showed that his bees had died of starvation. They still had plenty of pollen to supply the small daily requirement necessary for the health of adult bees but no honey—absolutely none at all. What is more, there was the telltale evidence of particles of wax on the floor of the hive showing that robber bees had been at work carrying off what scanty stores there had been at the beginning of the winter season.

"What a sad sight," Lee commented wistfully. "I wish I could have saved them."

"I know how you feel," I agreed, "for it's happened to all of us."

Usually, in our Santa Cruz area, bees do not swarm after August 1. But Lee had found the above mentioned bees as a tiny afterswarm the first week in September. He said the whole swarm, clinging to a bush, was no larger than a grapefruit. He had hived this little swarm into a standard hive body in which he had placed two fully drawn frames filled with sealed honey as the two center frames in the brood chamber, and then had placed four frames with starter sheets on each side of the sealed honey frames. Then he hived the bees and of course they made their home between the two drawn combs. Due to the fact that he acquired this swarm so very late in our honey season with the nectar flow practically at an end except for an occasional ornamental red blooming eucalyptus tree in someone's front or backyard, he was immediately confronted with the problem of robber bees invading his new beehive. He restricted the entrance with little wooden cleats until it was only one-half inch high and one inch long. But even that small opening was too large for his tiny swarm to guard, especially on warm days when his little force of workers sent out as many fielders as possible to gather both nectar and pollen. So he closed the entrance even more until he thought he had solved the problem. This hive was at his sister's place across town from his home so he could observe it at only three or four day intervals and often when the sun was not shining. This deprived him of a chance to count the bees leaving or returning to the hive, an absolute necessity when trying to bring a small swarm through the winter.

In my experience he made a mistake right in the beginning when he placed his two fully drawn and sealed combs in the center of a standard ten-frame hive body. A grapefruit sized swarm of bees simply does not contain enough bees to guard all four sides of such large centrally located honey-filled frames. This allows determined robber bees to obtain just enough honey from time to time to encourage them to continue trying to rob out the hive no matter how small one makes the opening hole. What he should have done was place the two drawn combs close up to the wall of the hive on the left side and then place one frame with a starter sheet next to those. Then he should have made a thin wooden divider board (one-quarter inch plywood would do in this particular instance) of a size large enough to completely separate the other seven frames from the three-frame brood nest. This procedure would have given his tiny swarm a sporting chance to defend its new home and at the same time allow the bees room to expand as much as they would be able to do in the fall and winter months. It would also have aided greatly in conserving heat, an absolute necessity for survival over winter. In the spring, about March 15 in our climate, he could have removed the divider board and let the bees continue to draw out the remaining combs. This method is much like the one I used in 1972 to bring through the first winter the tiny swarm I found in the garden which ultimately in 1974 became the world's record producer of wild flower honey, *Guinness Book of World Records* 1976–1979 under the heading of Insects and discussed fully in my book *The Art & Adventure of Beekeeping*.

There's a Bee on You

Honeybees become tired as they fly and need to stop occasionally to rest their wings. This is especially true during our sunny but cool winter weather. Often as we stand talking to a visitor in our backyard a homecoming bee will alight upon one of us. A stranger to the ways of bees sometimes becomes alarmed but there is no danger whatever. Such weary resting bees never have the slightest intention of stinging. All they want to do is find a warm, comfortable place upon which to land and rest their wings and catch their breath before resuming their flight.

A tired bee may stop almost anywhere. Quite often I have had one alight on the lobe of my ear and after a short rest it became interested in its surroundings and proceeded to explore the other parts of my ear which to a bee must be a singularly large warm cavern with interesting possibilities. At any rate I have had bees crawl slowly and carefully all around the perimeter of my ear and then on into it so far as to even alarm me though I know from long experience that there is not the slightest danger of being stung as long as I allow the bee to continue with its investigation without interference on my part. Of course, if I attempt to scratch a bee out of my ear or knock it off with a swift brush of my hand I may run some risk of being stung, not because the bee wanted to sting but because I frightened it at its work—trying to find nectar or pollen in my ear. I suppose a bee thinks that a large flower-like appendage like a human ear would surely have something of value to a bee colony necessitating the desirability of a thorough examination of the premises which may take as long as five minutes to complete. But believe me, it is a truly spooky feeling for those of us being so examined!

There's a Bee in the House

There are times when on my way home from our shop in the barn a weary bee will alight on my back and, all unknown to me or to the bee, will take a free ride with me into my house. Then as the bee becomes rested and tries to return to its hive it finds itself in strange surroundings. It will invariably fly toward a lighted window in the daytime and try to fly through the glass. Not being successful in this endeavor the poor bee will feel trapped and will buzz up and down the windowpane in a desperate effort to escape. When it becomes tired it will crawl up and down and all around on the vertical pane of glass still searching for a way out. 🐝 As soon as I hear a bee buzzing at a window I take a water glass and place its open end over the bee and against the window. A stiff piece of paper or an old envelope slipped between the open end of the water glass and the windowpane effectively traps the bee and I can carry it out of doors again and release it to fly away to its hive. But if no water glass is handy and the bee is trapped against the plate glass window in a store or in a church I can carefully surround the buzzing bee with my bare hands and slowly close my fingers as though making a snowball and the bee will stop buzzing almost as soon as the cavity made by my hands becomes dark and I can carry it outside to freedom. No bee has ever stung me while in the act of making such a rescue. A neighbor lady, Mathilde Sivertsen who used to almost panic when a honeybee came near her, recently was equally successful in barehandedly saving the life of a bee that was in imminent danger of being killed by the shopkeeper as it flew against the window of a downtown store.

Not long ago on a Sunday evening a honeybee flew into the lighted foyer of our Advent Christian Church just as we were all starting to leave for home. Someone called my attention to the bee as it clung to the wall. I tried to enclose it in my hands but before I could close my fingers it slipped out and flew first to the ceiling light and when we turned out that light it landed on the cement sidewalk under the porch light. Pastor Bill Creecy pointed it out to me and this time I caught it in my hands and held it up to the ear of Mildred Aaron, a blind lady in our congregation. She was delighted as she heard the sound of the bee buzzing for a moment or two before the darkness caused it to stop buzzing and start crawling around in the palm of my hand. Almost everyone present expressed surprise that I could pick up a honeybee in such a quick and easy way. I think some of them definitely expected me to get stung but the little bee quieted down and I carried her around the corner of the church to a place where it was dark and put her down on the leaf of a shrub. The night was relatively warm so I knew she could stay out all night and fly home when the sun came up the next morning. A beekeeper with love in his heart is always ready to rescue a beloved honeybee. And you can do it too!

3

Bees As A Business

Tips on Buying Bees

IN MY FIRST BOOK, *The Art & Adventure of Beekeeping*, I tell how we build "catcher hives" which induce swarms of bees to come to us of their own accord and occupy a hive that we have prepared and set up on the low roof of a building; or better yet, a hive set upon a shelf mounted very securely and without any wobble close up under the eaves of our house. But in these later years so many folks are on the lookout for swarms that even with our background of experience we have not been too successful in acquiring many hives in this manner. The reason is that most swarms cluster at least once somewhere near their own home or hive before they take off for our more distant catcher hive and this gives both amateur and professional bee-keepers more than a sporting chance to hive the swarm before it can come to us. But even this situation has its good points because more and more people are going into beekeeping with the result that there are more swarms available in many areas and some folks, both men and women, hive these swarms with the purpose of keeping them for a season or two and then selling them. I heartily approve of such action, for these bee-keepers when they sell a hive or two, always place a very fair price upon their bees and equipment and the buyer gets a good deal.

In our own Santa Cruz area I act as a go-between for both the seller and buyer of bees. I do this as a free service. Someone who wants to sell a hive telephones me as do those who want to buy bees. Often I can make a quick deal for both parties. A case in point is that of Mrs. Brunelle. She came to see me and said that she and her husband and three young sons had to move from their rented farm home and vacate the property in just seven days, and they needed to sell their two hives of bees as they did not know where they were going to move and so could not take their bees with them.

"Can you help us?" she asked. "We don't have time to advertise our bees and equipment for sale in any regular way."

"Yes," I said. "What do you have to sell?"

"I have it all written down on this paper," and she handed me a neat slip of paper with the following written upon it:

FOR SALE: Two good-natured working beehives and equipment.
A. One hive consists of two brood boxes and two honey supers.
B. The other hive consists of two brood boxes and one honey super.
Also included: one hive tool, two queen excluders, one smoker, two well-made bee veils, 12 honey foundation frames, entrance cleats.
The two hives and equipment complete—$50.00.
Albert Brunelle
Soquel, California.

"Do you really want to sell *two* hives and all that equipment for only fifty dollars?" I asked. "That seems like a very low price and you can surely get more."

"We could if we had more time, but we need to sell right away. And you know that two years ago we bought an empty

hive and super from you and have been making almost all of our equipment ever since including our bee veils. We figure we've fifty dollars in cash in these two hives and if we can sell for that much we'll be satisfied as we've also had a lot of good honey. And some day we'll have bees again, of course."

"Come in the house a few minutes while I check my records. I think I know a man who'll buy your bees."

Some months earlier Roger Ghiotti had contacted me wanting to hive swarms of bees and he had asked me to let him know of any swarms that might be called to my attention and I did not have the time to hive. I had been glad to take down his telephone number which I now gave to Mrs. Brunelle.

"Call this number when you get home," I suggested, "and let me know how you make out."

She smiled as she thanked me and left for home. The next day she phoned and said that she had contacted Mr. Ghiotti the previous evening and that he and his family had immediately come over to see the bees, had liked them and purchased them, and also all of the equipment. So it was a quick deal that turned out very well indeed and I was glad to have had a part in making the arrangements. Every beekeeper should be ready and willing to help another beekeeper in every way that he can for our honeybees need our united assistance.

Setting up Shop

Ken had come to see me several times in the fall of 1975. He had been reading up on beekeeping and then when my book came out he had purchased a copy and read it through. More than ever he was now convinced that he should go into bee-keeping full time. He had sold a larger property and had the money to make the initial investment required for buying bees and equipment and still have enough left to support his family for one year. However, he definitely would need income from his venture the second year, and preferably realize some return, even on a small scale, the first year. I liked Ken and his down-to-earth ideas so as the days passed I had several hour-long discussions with him covering many facets of the beekeeping industry. If one owns two to five hives of bees he is a hobby or backyard beekeeper. If he owns as many as twenty hives he is entering the small business category. And if he owns as many as 100 or 200 hives, or even more, he is in the beekeeping business and must make the business pay for he will have many expenses, some of them quite unexpected.

Equipment Needs and Decisions

All who get into beekeeping as a hobby or a business know that the first two years one is always spending money for it is necessary to buy and build a backlog of supers to be placed on the hives as needed—and one never seems to have enough during those first years. Whether one buys supers knocked-down, completely assembled, or builds them himself it seems that one is always a few short of the immediate needs as long as there is a honey flow. At this season the beekeeper is kept hopping to try to keep up with everything that needs doing. And then during the latter part of the honey flow and thereafter one needs to buy or build an extractor room and all related equipment to extract and pack the season's honey. My father and I like a permanent extractor room but excellent portable extractor rooms mounted on trucks or trailers are available. But again, they cost money. However, if one's apiary is some distance from home such a portable extracting set-up is a great advantage. Supers can be extracted early in the afternoon and at least some of them replaced on the hives later the same day and one does not have to haul supers home, extract them, and haul them back to the outyard the next day.

Record Keeping a Must

Keeping the necessary records is a continuing problem for all of us. But if we want to try for a world's record it is an absolute necessity to keep an accurate record of each hive both for our own use and also to submit as proof to Guinness of London in the event that we succeed.

Bees As A Business

On my kitchen wall I hang a large free savings and loan calendar that has spaces for the days of the month marked off in squares with a small date number in the upper left hand corner. Thus I have room to make entries as required for those hives that I am particularly interested in for maximum honey production. I make the entries for lesser hives in a small notebook. I always make an entry on my calendar or in my notebook when I place a super on a hive and jot down the hive number. Such a calendar can accommodate the entries of ten to a dozen hives because one seldom supers all of the hives on the same day. Every day as I walk through the kitchen and glance at the calendar I am reminded of my bees and their urgent or coming needs.

Later in the season when I take off a filled super of honey I make an entry in ink on a one-eighth inch thick piece of hardwood plywood that is about one foot wide and two feet long. On this board I have an entry for date, hive number, and weight of honey extracted, as follows:

	Hive Number	Honey extracted, lbs.
May 3	4	32
28	8	28
29	3	29
June 2	4	61

I like this type of record because it gives me a visual summary of what individual hives are doing as well as the overall production of my apiary.

In California it is now required that a producer pay an assessment of ten cents for each sixty pounds of extracted honey produced, and twenty cents if a beekeeper is both a producer and seller of his own honey. The marketing year begins on April 1 of each year and ends on March 31 the following year. The assessment money due must be sent to the Department of Food and Agriculture, Emil M. Loe, Chief, Bureau of Marketing,

1220 N Street, Sacramento, California, 95814, on or before March 31 of each year. Check with your State Department of Agriculture to see if any such requirement applies to beekeepers in your area.

Guinness

When one submits records to Guinness it is necessary to make a compilation of supers placed, taken from the calendar; and a second compilation of supers removed, taken from the plyboard record. Photographs and other information are also required.

Marketing

A beekeeper can usually sell up to 1,000 pounds of honey from his own home as friends and neighbors buy honey and pass the word around to others. But beyond that amount it becomes more difficult to dispose of a bountiful golden harvest. I like to do a certain amount of peddling but usually a better way these days is to contact health food stores and farmer's markets. The most profitable way, if one can produce much honey and there are several members in the family who can help with the work, is to make arrangements to set up a roadside stand on a well traveled road or highway. People love to stop at such a stand and they will often pay almost double the usual price, if necessary, to secure honey sold along the road. Many people just want to try the local honey from every area they pass through. Honey packed in small glass containers weighing from a few ounces up to three pounds sells well in these roadside stands.

Choosing a Home Apiary Location

Those who have a family and are considering beekeeping as a full time occupation must give serious thought to the matter of adequate housing, schools for the children, and a church for the whole family to attend. In my experience regular church attendance is an absolute necessity for successful beekeeping. We are dealing with basic creatures of God's creation and we need His help and guidance to maximize our chances for success.

And out here in the West where there are still great wide open spaces it is sometimes many miles between towns where most of the schools and churches are located. Long bus rides to school can be very tiring for children. I know—for I had to endure them.

Evaluating Bees for Purchase

On November 19, 1975 Ken again stopped to see me.

"I've definitely decided to go into beekeeping," he informed me. "I have an option to buy 200 hives for fifty dollars a hive from a man near Fresno, California. What do you think of that? Will you go with me to check out the bees when I go to inspect them on December 4? I'll pay you to go with me."

"I can't go with you on that date," I said, "and as a matter of fact you don't need me for you yourself can do everything that I could do. Let's talk it over and I'll give you some tips on how my father and I check out bees offered for sale."

I then told him that we consider the fall of the year, at least in our Santa Cruz, California area, as being the worst possible time of the year to either buy or sell bees. Personally I would neither buy nor sell bees at this time for the risk is too great that some of the hives may winter kill, oftentimes for reasons unknown, but our average is the loss of one hive in five. We sometimes guarantee our bees to come through the winter active and strong. But if we sell all of our excess hives in the fall we have none to replace possible loss sustained by a buyer during the winter months, and that loss is always an uncertain factor as some years we lose a greater percentage of hives than in other years. So we do not plan to either buy or sell bees until sometime after March 1. However, sometimes one has no choice in the matter. On occasion we have had to move in the fall of the year and could not take all of our bees with us. Then we had to sell some of them. I always explained to the buyer the risks involved, the precautions he could use to minimize his losses, and also offered to sell the hives at a lesser price than they would bring as of March 1 to compensate for the lack of a guarantee, since the buyer was taking his own chances. Invariably the buyer accepted my proposition and acquired the bees.

🐝 Sometimes a person must buy bees when he can find them for sale no matter what the time of year. To reduce the risk one should follow a few simple rules. First, before going to the expense of making a long trip, write to the bee inspector of the county where the bees are located. In our county he is listed under County of Santa Cruz, office of Agricultural Commissioner. Ask him when the bees offered for sale were last inspected and the reputation of the seller. Then check with the local bee inspector. He may already know all about those particular bees if they have been for sale for some time. He may also know the current selling price of bees from that area, or he may be able to find out as bee inspectors often work together and discuss mutual problems such as disease, stolen bees, bee equipment and hives.

Examining the Hives

If these preliminary investigations check out to our satisfaction we make an appointment to see the bees. Upon arrival at the seller's bee yard we quickly scan the arrangement of the hives. 🐝 If some of the hives lean forward, others backward, or are about to fall over sideways you know that the bees have had little recent care. A neat placement indicates a practical and probably trustworthy owner. But we do not, just for that reason, take his word for it that all the hives are strong and worth the asking price. We take time to carefully check out each hive. If we have to drive a considerable distance to inspect the hives to be purchased we may find that on arrival the weather, in late fall and early winter, is overcast, cold, rainy, or clear and frosty. In all these situations one will see no bees flying—none at all. How then can one tell if the hives contain bees or are just a row of empty boxes? I kneel down by each hive, place an ear up against its side, and listen. 🐝 A good strong hum from each side indicates a strong colony and probably a good one. A strong hum from only one side indicates a weaker colony but one that might yet survive the winter very nicely. The weaker the hum the greater the risk to the purchaser. Even on a cold, thirty to fifty degrees, or rainy day when no bees are flying there should be some sound of humming going on in the hive. If the temperature is in this range and one hears no sounds whatever, that hive is suspect and a written note should be made of it so that, if possible, an inspection can be made of it at a later and warmer date. Of course if the weather is so cold that the temperature is down in the twenties or lower there will be no appreciable activity in any hive and listening will not help in ascertaining its current condition. The colder the weather the less activity in the hive until it ceases altogether.

Sea Shell Effect

An empty beehive or even one with all ten frames in place but without any bees in it at all is never completely silent as one presses an ear up against the side to listen unless the entrance is tightly closed with a piece of wood. An even then there is usually some sound to be heard at least during the daytime. The reason is that an empty beehive resounds to all the noise, street traffic, voices, and other sounds that are being made in the neighborhood. My father and I call this the "sea shell" effect for it is much like that of a large conch shell held close to one's ear which seems to murmur the sound of waves breaking on the seashore. In the usual course of events one need pay no attention to this phenomenon but if one is watching the scout bees around a catcher hive or even an ordinary empty hive setting on a stand one may be deceived into thinking a swarm has taken possession when in reality there are no bees in the box. Sometimes I have seen several dozen scout bees around a hive box when I had to leave in the morning and when I came home at night and listened to the side of the hive and heard the roar inside I was delighted until I had listened a short time and had detected the slight difference in the sound between the sea shell effect and the presence of bees in a hive. There is a difference, but the only satisfactory way to determine it that I have found is to set up an empty hive early in the spring and with the entrance completely open, listen at various times during the day and night to familiarize myself with the sound. Then when a swarm of bees does take possession and I place an ear against the side of the hive I can instantly tell that bees are actually in possession. This is always, of course, a great joy to a beekeeper.

Again, if one is buying bees and must make his inspection on a cold, rainy, or foggy day, or in the evening, it is a great help to know the exact sound of bees in a hive. One does not want to waste time considering the purchase of an empty hive no matter how interesting a hum it may have.

When my father and I go bee buying it is my job to listen to the bees humming in their hive while he stoops down and smells a few good whiffs at the entrance. He has an excellent sense of smell far surpassing mine. If he detects even a faint odor of rottenness he touches me on the shoulder and I make a note of the hive number and location. Then I too take a sniff and if I can scent even a trace of rotten-meat odor we definitely decline to purchase that hive either at the moment or later. My father's ability to quickly detect the telltale odor of what might well be the beginning of foul brood or some other malady peculiar to bees is probably part of the reason we have never had disease among our bees.

Weighing a Hive

Next, it is important to obtain a reasonable estimate of the amount of honey in each hive that the bees can use as winter stores. My father and I use a fish-type spring scales capable of weighing up to one hundred pounds to make this estimation, although two fifty pound scales in tandem or coupled side by side will do equally well. For lack of something better, in an emergency one can use a bathroom scales by standing on it and lifting up on the back end of a beehive. Subtracting one's own weight from the total will give the weight of the hive's rear end. We usually hook a chain from the scales to the bottom of each hive and weigh the back end only. The rear end of a single

Weighing A Hive

depth brood chamber as of December 1, in our area, should weigh approximately thirty pounds, a double depth brood chamber should weigh forty-two pounds. A few pounds greater weight is preferable. A single depth brood chamber rear end weight of only eighteen pounds means that one will surely need to feed that hive before spring or the bees will starve. Less than sixteen pounds usually means that the hive is empty and one may remove the cover to inspect the interior regardless of the time of year or the state of the weather. ❧ But do not buy the hive without opening it, thinking that at least you are acquiring some good equipment, for there have been instances where all of the frames have been removed and the cover replaced. Someone had removed the drawn combs and, everything looking natural, the owner was not aware of the missing frames. Such stripped and empty hive boxes may have but little value. I should state here that when we weigh the rear end of hives we are taking it for granted that the hive being weighed has the usual short horizontal entrance board projecting approximately three inches beyond the front of the hive itself. If the hive bottom should by chance have a longer or shorter bottom board one needs to make allowance in his weight table to compensate for the difference, but such odd length bottom boards are rather rare these days. All that one needs to remember is that the fulcrum point should be three inches forward of the front of the hive. Various woods used in the construction of hives do vary in weight so as the season advances and fall comes on, each beekeeper should compile his own weight table for future reference. Ample stores of honey in every hive is also excellent insurance against disease.

❧ If we can tell by the hum that there are bees in a hive we never remove the cover from the hive in late fall or winter to inspect the hive for bees or stores. Doing so breaks the propolis seal thus releasing warm air all the balance of the winter. To compensate for the heat loss the bees are compelled to eat more honey and by doing so may use up all their stores and starve before spring, whereas if the cover had not been removed they would have had ample honey to carry them through.

Winter Feeding

Occasionally in winter upon opening a very light hive, we have found a good nucleus of bees. If the brood frames were only partially drawn the season before and the bees had no opportunity to store much honey or pollen the weight of the hive will be accordingly less; and as of December 1 there may still be enough bees in the hive to carry them through the winter provided we give them enough honey to supply them with food. 🐝 In such a situation we always feed pure honey as sugar water or other substitutes tend to further weaken an already weak colony. As a matter of fact my father and I always feed honey, if feeding is necessary, and we always store away enough jars of pure honey to feed all of our hives in the event of an exceptionally cold or long winter. As a rule we have a considerable carry over of honey but that is good for there is always a ready market for finely crystalized honey in early spring. Some of our customers really prize such honey.

Winter Feeding and Protection

In the event that we have made an error in our appraisal of a lightweight hive and, upon opening, have found bees in it, we make a hasty inspection and replace the cover to conserve heat. 🐝 As soon as practicable we add an insulating cover to conserve additional heat and then wrap the hive with four or five thicknesses of new black plastic sheeting, draping the plastic down on all sides until just above the entrance. We tie it in place with two heavy strings or twine that completely encircle the hive. The black plastic absorbs heat from the winter sun thus aiding the bees in keeping the brood chamber warm which in turn helps conserve the additional stores we give the bees. At intervals of every other day we give such a hive a feeding of five or six tablespoonsful of honey in an entrance feeder. 🐝 Our usual feeder is a small homemade rectangular pan with a wooden rim one-quarter inch high, a width of three inches and a length of six inches. The pan has a thin sheet aluminum bottom. I seal the wooden rim to the bottom with beeswax so that it will not leak the warmed honey I give the bees. The addition of a little float made of wooden slats with spaces between completes the feeder.

Caution: Sometimes when proper insulating cover material (soft fiberboard such as celotex) is not readily available people ask me if they can use several thicknesses of burlap gunny sacks as a substitute. Yes, but not if you intend to cover the whole hive with plastic unless you thoroughly wash and then rinse in several waters the sacks you intend to use. There is some evidence that on occasion certain feeds are bagged in sacks that have been treated with a poisonous chemical to repel weevils and other pests that would destroy the feed in the bag before it can be sold and used. This is probably an ideal way to preserve feeds from insects and rats, but as beekeepers we need to take proper precautions when we use these sacks for purposes other than that for which they were intended. Smoker fuel sacks should be thoroughly washed also.

Buy the Land

In this present age of ever rising prices for almost everything we purchase including bees, beehive equipment, extractors, trucks, gasoline, and land, it is imperative that one either rent on a longer term basis (three or four years) open space land upon which to place his purchased bees—or better yet—buy it outright. As a matter of fact through the many years that my family has owned bees, we have had the opportunity to realize more money from buying and selling the land upon which we placed our bees than we have from the honey we sold. In my lifelong desire to set a new world honey production record I have sometimes urged that we move before our property had a chance to rise in value. Again, there have been times when we wanted to sell and the purchaser did not have sufficient funds to pay the full market price. Such a situation developed when we wanted to sell out a few years ago and move here to our present location. A young couple wanted to buy. They had a baby son and needed a home. They had also bought honey from us for some years and greatly desired to learn beekeeping. We all liked this young couple and I recognized in them a free spirit like myself who loves God's great out-of-doors and feels depressed and frustrated from the day in and day out drudgery of a continuous indoor existence. But these dear young folk did not have the money to buy our place. Even with as large a loan as they could swing they did not have sufficient funds. They appealed to their parents who came to look at our place and liked it but even with their help—and parent's help is often needed and most appreciated these days—they still could not pay what should have been the current selling price. So my mother, father, and I talked it over and we lowered the selling price to the point where our young friends could afford to buy it. The passing years have proved that we did right. We still eat well and often and they do too—and their three little boys have a parcel of land on which to live and grow. Recently I was questioned by the young father as to whether I was sorry we had sold to them at the price we did.

"Definitely not," I answered with a smile. "You needed a home at that time and you still do." He laughed and thanked me again.

About this time a lady queried me on the same subject. "You could have sold for far more than you did," she chided me.

"Yes," I answered, "but they couldn't afford to pay any more, so we sold at a price they could pay." She shook her head.

If during the years an acreage looked good to us for placing bees we knew that sometime in the future it would also look good to someone else. A deep wooded gulch away back from a main road, even though in close proximity to a town, was not formerly considered of much value. Yet today such land brings a high price because people are desirous to get away from it all, at least as far as they can reasonably get, and they like these backwoodsy locations. Now and then one can still find such a property at a reasonable price.

One may look at bees and consider buying a few hundred hives but before actually making the purchase one must *rent* or *buy* the land upon which to place them. I cannot stress this fact too strongly. To make the honey business pay there are many factors to take into consideration. 🐝 Two hundred hives require at least ten acres of land with no near neighbors (one hundred acres would be far more preferable) and land that is accessible by truck and yet far enough away from a public roadway that it will not constitute a hazard to pedestrians or vehicles. Also it must be located away from the homes, stables, or playgrounds of any neighbors in the area by approximately one-quarter to three-quarters of a mile depending upon the lay of the land. An intervening hill is excellent. Even a strip of tall timber in between helps. If your neighbors are also beekeepers you will work more or less closely with them, and they, not being afraid of your bees, will enable you to place your hives to greater advantage.

Bees do swarm, particularly for the beginning beekeeper, and 200 hives purchased in spring or early summer, may by late summer have increased to 400 if the owner has had the foresight and funds to either buy or build the additional 200 empty hives needed to catch and hive most of his swarms. There is no time to buy or assemble new hives after the swarming season has begun. All such preparations must be made in advance. And after hiving his new swarms the beekeeper must either find a buyer immediately for these new hives or rent or buy more land in another location.

🐝 Two hundred hives, as a rule, is the greatest number that may be placed in one location to obtain maximum honey production. Often people are unaware that the buyer of 200 hives must be prepared that same season to either build or purchase a great deal of additional equipment to hive new swarms. To just let the bees swarm and fly away is, of course, a great waste of one's resources. It would be far better to buy 100 hives in the beginning and increase to 200 by natural swarming, and take good care of all of them, than to buy a larger number and then be unable to adequately handle the natural increase.

Basically, 200 hives require much open-space land away from subdivisions, suburbs, small towns, shopping centers, public gathering places, and the like. Through the years many lawsuits have been brought by neighbors and others against the keepers of bees. But unless the complaining party could show true cause he lost the suit every time. He must show actual damage caused by the bees, or suffering directly caused by them, usually on a continuing basis. Just because an uninformed, fearful, or troublesome neighbor or official makes up his mind that he is going to get rid of a certain beekeeper because he does not like either the beekeeper or his bees does not constitute legal grounds for his being able to do so. Nonetheless, as beekeepers, we must be prepared to go more than half way to keep peace in our area though there is a limit to which we can be pushed around by unreasonable persons. Honey is a vital and valuable commodity that goes largely to waste and we are rapidly approaching a period in our world's history where such waste cannot be tolerated even to satisfy the whims of a considerable minority. I am happy to say that of late years the television industry has done a commendable job of helping educate the general public by showing numerous fine films on bees and beekeeping. Newspapers and magazines also merit our commendation and thanks.

Finding a Profitable Apiary Location

Along the Pacific coastline from Mexico to Alaska there are still some good places to locate bees that are amply removed from human habitation to permit the location of from twenty to two hundred hives. With the present price of honey and beeswax one does not need as many hives to make a fair living as was the case even a few years ago. Finding the best possible location does take time and gasoline. But finding a reasonably suitable location can be done in a few months of search and inquiry. Almost anywhere along the Pacific coast where one can find enough open space to keep bees, one should do quite well with proper handling of his hives. Even south along the Monterey coastline, where there is often fog in the winter and spring months, there is excellent pasturage for bees. Oldtimers used to tell me that right along the coastline bees did not do well because the climate was too cold and damp during the season of nectar flow for the bees to fly to advantage. There were too many days when the bees were unable to fly at all. But now with improved hybrid bees such as Ames Apiaries of Arroyo Grande sell, bees that can work in cooler weather than our oldtime Italians, the outlook has changed. And I think the weather has become more mild and sunny too, at least for the past few years.

A friend of mine, Edna Solari, told me that more than one hundred years ago her grandfather, who loved bees, rode horseback along the rugged and wild coastline south of the city of Monterey, and thirty miles or so along the way he saw a particularly beautiful little valley among the towering mountains. He bought two acres of land for when he retired he wanted to settle there and keep bees. But though his beekeeping plans failed yet he continued to hold the land and today his little dream apiary site is still part of the family property. After hearing my friend's story my father and I went for a look to see what her grandfather had found in that early day that he thought was better than any other place he had seen. When we found it we were as enchanted with it as he had been! Upon closer inspection, after a considerable walk up and down a mountainside, we found that someone else had already capitalized upon the same idea we had in mind—there were about eighty neatly cared for hives in a thirty acre meadow on the floor of the little valley. The bees were working lustily and it was a joy to watch them. They were in an ideal location in that they were on the valley floor with moderate to steep mountainsides just covered with innumerable bloom on every side. The bees could fly uphill to get their loads and then coast downhill with their loads to the hives. There was not much wind, either, though much of that country has strong prevailing winds blowing inland from the Pacific Ocean as the tortured shapes of many trees testify.

Not far beyond that point is located the boundary of Fort Hunter Liggett, formerly called the Hunter Liggett Military Reservation. The actual military base is situated many miles away on the other side of the towering Santa Lucia range of mountains. Whether or not one could get permission to place beehives on any part of the reservation, particularly the seaward side, I do not know. Surely there are many ideal spots. Permission granted or refused would probably depend upon whether our country was in a time of peace or war. Farther on one finds some private, though high priced, land suitable for beekeeping and then comes the great Los Padres National Forest. I would surely like some day to place a few hives of bees a couple of hundred yards behind the forest headquarters buildings. It appears to be an ideal site. I looked over the area three years ago when I drove down there to talk to the chief ranger about the possibilities of placing hives on National Forest land, but he had been called away for the day so I missed him. That spring our own No. 4 hive showed every indication of setting a new world's production record for wild flower honey, which indeed it did, and we have been too busy to go down again to see the ranger. There are many National Forests and most of them that I have seen have good honey producing areas for bees.

The last week in July 1977 I wrote to Mr. Robert E. Breazeale, District Ranger, United States Department of Agriculture, Forest Service, Los Padres National Forest, 406 South Mildred Avenue, King City, California 93930, asking him about the possibilities of placing beehives on National Forest Service lands in his area. Before he had time to more than read my letter, lightning started the great Marble Cone forest fire in the Los Padres National Forest which burned furiously over a vast area. I wondered if he would ever have a chance to answer my letter. Just now, to my delight, I have received an answer from him. His informative letter reads as follows:

October 11, 1977

Dear Mr. Aebi:

Thank you for your inquiry concerning obtaining a permit for establishing apiaries on National Forest land. I apologize for the delay in responding to your inquiry. As you may already know the Marble-Cone fire started on August 1st and suppression activities lasted nearly a month. Since then we have been preoccupied with rehabilitation work and catching up on our normal District work.

Apiaries can be allowed on National Forest lands and are authorized by issuance of a Special Use Permit. Application is made through the District Office that administers the land where the applicant has selected a proposed location. Upon receipt of the completed application the Forest Officer in charge reviews the application, the site selected, other uses of the area and future plans for the area that may be affected by the proposed apiary. Upon completion of the review, an environmental analysis report is prepared with recommendations on issuance of the permit.

Specifically, our main concerns with regard to apiaries are:
1. That issuance of the permit will not allow conditions which conflict with State or local ordinances, Laws have been designed to control apiaries and prevent the spread of diseases affecting bees.

2. Locations will not be approved which result in conflicts with other National Forest uses. For example, recreation areas and stock driveways will be avoided.

3. The number of hives that can be successfully worked from a site depends on the feed available in the surrounding area.

The permit is subject to charge. The minimum fee is $20.00 per permit for up to 100 hives. Additional hives are $.20 each.

I have enclosed a Special Use Permit application for your information or use. If you have further questions please contact Peggy Robishaw of my staff at 408-385-5434.

Sincerely,

ROBERT E. BREAZEALE
District Ranger

Hazardous Locations

A good many years ago we took a trip through the Mendocino National Forest. Away up near the top of the mountain just beyond the ranger station we passed half a dozen beehives at a little distance from the road. We stopped and walked over to inspect them more closely. There are bears in that country and bears love bees and honey. I wondered if the beekeeper was aware of the bears. Indeed he was, for he had built an electric fence composed of many charged wires up to a height of almost eight feet all round his hives. The fence was powered by a battery and every few seconds we could hear a click denoting that the power was on. That man was taking no chances. He knew the risks and had taken the precaution to reduce them to a minimum. Sudden storms, animals, thieves, poisonous insecticides and vandals are the great destroyers of our friends the bees. Excellent government inspection of beehives has largely eradicated disease among bees. The drug companies have also been of great assistance in controlling disease. Some of the drugs almost work miracles.

In looking for an apiary location one must always take into consideration the hazards involved. For instance, when locating bees inland in the foothills surrounding the Sacramento and San Joaquin valleys one finds buckeye trees. These trees bloom copiously rather late in the spring and early summer and usually bloom for about six weeks. They provide both nectar and pollen for bees. For many years it was thought that the nectar was poisonous to bees but recent experiments by Mr. Al Shatz of Concord, California have shown that it is only the pollen that poisons the young brood. 🐝 When he uses a pollen trap to exclude buckeye pollen from his hives all of his bees thrive.

My father and I have kept bees in areas where there were occasional buckeye trees and we noted no ill effects whatever from the buckeye bloom even though we did not use pollen traps. I think this was because there were still some eucalyptus trees in bloom in our area. Being intelligent little creatures the bees probably knew that it was risky gathering pollen from the buckeye trees and since they had a choice they left the buckeye untouched, at least I never saw any bees, either ours or those of anyone else, working on the nearby buckeye. Yet that same year we drove to a large area of blooming buckeye and saw myriads of bees working on it. Apparently if bees cannot find any other bloom they will really work on buckeye, being hard pressed to find something to feed their hungry brood, and they suffer as a consequence. This past summer, Stan Phillips, a friend of mine living a few miles from me in Ben Lomond, said that suddenly one hive of his bees began working on a buckeye tree and soon thousands of sick larvae were being dragged out of the hive. Fortunately for him the buckeye blooming season was near its end and his hive survived. He said his other hive was in no way affected. He thought they must have had an entirely different source of nectar, or they did not have as much brood to feed. It was his strongest hive that was so adversely affected.

Again, when looking for a profitable apiary location for the early honey flow, we must keep in mind that in much of this country we have high mountains, sudden storms, and flash floods. Since it is usually advisable to place beehives so that they are protected from strong winds, and dry washes are usually fairly level and commodious, it is a temptation to place bees in such seemingly ideal locations. But we never do that because, for reasons we cannot ascertain so early in the season, we may not be able to move our bees for a year or even two years. In the meantime a flash flood can wipe us out. All down through the years my father and I have seen the wreckage of beehives strewn along various waterways. It is a most disheartening sight, and it meant ruin for some beekeeper.

Some years ago we heard over the radio that there was flooding along the Sacramento River. About ten days later after the weather had cleared and the waters receded we drove up to the vicinity of Rio Vista in the delta area. The river had returned to its normal size but all up and down the levee, scattered helter-skelter, were parts of beehives—dozens of them. Hive tops, hive bottoms, brood chambers, and honey supers were lodged in the willows or reeds for more than half a mile where we walked along on the river side of the levee. Some of the brood chambers and full depth supers were still in good condition in that they were not broken to pieces, but the whole interiors were filled solidly with silt that had been washed in between the combs. I was curious to see what damage the flood had done to the combs and if there had been any larvae in the combs when the flood struck. So after much effort, for the mud made the boxes very heavy, my father and I managed to drag two parts, a brood chamber and a super, up the levee bank and loaded them into our car to take home and wash out. The silt washed out relatively quickly and easily with the jet from a hose and I was surprised to find how little damage had actually been done to the wax combs. Except for one place where it seemed a tree limb had poked down between two of the frames, thereby ruining two combs, there was no serious damage. There was sealed brood in some of the frames, dead of course, but not in a stage of putrefaction. Apparently the cold mud had preserved them for the ten days since the flood. Most of the honey was still sealed in the combs and I was tempted to eat a little but thought better of it as I could think of no way to sterilize the honeycomb from the polluted river water. But after washing it looked good enough to eat.

After thoroughly washing the brood chamber and super I was tempted to keep them and the next spring hive a new swarm of bees in them. But since my father and I have never had any disease among our bees I decided not to keep the equipment but burn it, as there was no way we could think of to test it for foul brood or any other disease. It simply was not worth the risk. But I would have liked to have known how the bees would take to living in what had been a thoroughly soaked and muddy hive, especially since they are such neat house-keepers.

Placing the Hives to Best Advantage

Now that we are living in the age of bulldozers a little expeditious use of such a machine, even if one has to hire the work done, can make a suitable location for placing bees even on sloping, gullied, or brush covered terrain which in years past would have been totally unsatisfactory. Careful advance planning before the bulldozer arrives always pays. In rugged terrain we select an apiary site high enough above the valley floor to enable the bees to utilize the early morning sun. When the bulldozer operator arrives we want him to cut two terraces in the side of the hill or mountain running in a north and south direction so that when our hives are in place each hive will have its entrance facing east, or at least southeast. This terrace should be cut as nearly on the level as possible and not less than eight feet wide for we need room to walk completely around each hive to facilitate servicing it.

The second terrace should be cut just in front of and four feet lower than the beehive terrace so that we can run our truck or other conveyance close in front of our beehives to aid in adding or taking off supers. Supers of honey are heavy and a portable lightweight wooden walkway spanning the short horizontal distance between the beehives and truck bed is a great labor saver. One should always plan his apiary layout so that he does not have to carry honey and equipment either uphill or downhill, and for as short a distance as possible. A beekeeper's days become long enough and hard enough without adding to the burden. Proper drainage to minimize erosion must always be provided and culverts or little bridges installed in the beginning insure a smooth roadway that helps to make beekeeping a pleasure and also more profitable. We do not want to needlessly bounce or irritate our industrious little friends when we move them.

Buy bees with care, place them with care, give them loving care, and they will care for their keeper.

4

Bees
In
Walls

THERE ARE MANY BUILDINGS in our Pacific Coast area that have bees living in them. This is due in large part to the fact that most of our homes and other structures are built of wood, and exterior wooden siding of all types has a propensity to shrink and swell because of variations in the moisture content of the air from season to season. The result is that as buildings age they develop cracks in their exterior walls and bees move in when they are in dire need of a home during the swarming season; and having once moved in they resist our efforts to remove them.

A Sticky Situation

April 4, 1977 a man named Charlie came to see me. He was the caretaker for the Batterson Nursing and Convalescent Hospital located only a mile or so from our home. He said that a swarm of bees had clustered on the limb of an apple tree on the grounds two days previously and then the day before it had moved and gone into a wall of the wash house. This was a separate little building close to the walkway between the main building and the parking lot. He was very eager to get rid of the bees and asked if I would consider getting them out of the wall. I really did not have the time to go but could find no one on my list of persons wanting bees that was free to hive such a swarm on a Monday afternoon. All of the names and information on my list said swarms wanted after work or on Saturdays. So after due consideration, and since there was urgent need, I told him that I would help him right after lunch.

My father offered to go with me and we drove to the place in a few minutes. The bees had entered the wall only about three feet above the ground at a point where the one and one-half inch drain pipe passed through the wall and down to the sewer. Whoever had installed the plumbing had cut a hole in the wall considerably larger than necessary to accommodate the pipe and the bees had made use of the crack around the pipe to enter the wall and establish their new home. I had offered to remove the bees for ten dollars provided we were able to get a good swarm of bees. But as I looked at the situation I did not like what I saw. The bees were very docile—entirely too much so. And there were some dead ones on the ground below the entrance. A close examination showed that they were not all dead either because some still moved a leg at intervals of a few seconds. I was puzzled. There was no way to account for the bees' behavior so I went ahead and set up my two sawhorses with a plank across them and set my beehive upon it with its entrance close up to the bees' entrance hole in the wall.

There is a way to smoke bees out of a wall if one has a certain amount of good fortune. We tried that method. It did not work. In fact we had no slightest promise of success. I felt even more frustrated when I found the wall was so constructed that the easy way to remove the bees should have worked. The bees were in the lower part of the wall below the diagonal two-by-four bracing that I knew was in the wall. The presence of such bracing was evidenced by a diagonal line of nail heads that otherwise would not have been there. 🐝 I had drilled a one-inch hole in the wall just below the diagonal bracing and another about a foot above the sole plate. Then I had used my big smoker (four-by-ten inch barrel) which was filled with burning burlap and forced a great volume of smoke into the upper hole to drive the bees down. After a few minutes I plugged that hole and forced more smoke into the lower hole to drive the bees up and out around their drain pipe entry. At the same time my father forced smoke into the crack under the pipe on the inside of the wall as an aid in keeping the bees moving. Our efforts caused a few dozen bees to hurry out of the crack to the outside and fresh air but not nearly as many as we had expected. I called to my father to come outside and we talked it over but were unable to come to any logical reason for the bees' failure to come out of the wall. There was no way the bees could get up into the attic, or go down under the floor. 🐝 If such unseen exit holes do exist our easy smoke out method will not work because the queen will flee into these inaccessible areas. Then the owner of the nursing home and others came by to see what we were doing.

"How are you succeeding?" the owner asked me.

"Not so good," I answered, "We'd hoped to get the bees out the easy way but they don't cooperate at all, so there's no easy way. We'll have to open the wall."

"Do what you have to do," she said, "but get those bees out—and save them if you can. We like bees too."

After she had walked away a short distance the thought came to me to talk to her again.

"Wait a minute," I said. "My candid opinion is that it would cost you less money if you hired a pest control concern to eliminate those bees. It's going to be difficult to remove the siding boards we'll have to take from the wall to get to the bees, and I think the boards will split so badly that you'll have to replace them with new lumber."

"That's all right," she said. "Do what's necessary. But what'll it cost?"

"Twenty-five dollars," I said, "it's become a big job now."

"Get them out," she nodded as she spoke. "I'll pay you that much."

So we went to work. I used a brace and three-quarter inch diameter wood boring bit to drill a hole through the siding close beside a stud, and then used a keyhole saw to cut straight down the studding the width of two boards or about fifteen inches. Then I used the saw to cut upward for the width of two boards. With a hive tool and a flat wrecking bar I tried to remove the siding boards without damaging them but they cracked and split in spite of my best efforts to take them off in one piece. The ends of the boards were less than three feet in length because the drain was that near to the corner of the building and I only had to take them off between two studs. But the redwood boards cracked all to kindling wood as I had feared they would.

Hundreds of bees flew all around me as I worked making what seemed to the spectators a most dangerous situation, yet not even one gave any indication of wanting to sting. After I had removed the four boards necessary to get a clear view of the bees I saw the reason. Thousands of dying bees filled the spaces between the studdings to a depth of more than eighteen inches! The bees were so exceptionally gentle because the guards were dead or dying and the queen was dead or dying too. Old honeycombs hung in festoons all the way up past where I had removed the boards and away on up out of sight. But there were practically no bees clinging to any of them. With bare hands I reached in and pulled some of the combs loose and removed them from the wall, placing them in a large cardboard carton that we had with us. Finding such a great amount of honeycomb in the wall surprised me but what surprised me even more was when I pulled down the second large piece of comb and a great shower of some kind of sticky liquid poured out of the cells and drenched my right hand. In a few moments my fingers and hand began to burn and swell. I dropped everything and ran to find a water hose so that I could wash my hand. I soon found one but the sticky stuff was difficult to wash off. Having removed as much as I could I went back to the bees in the wall. It was obvious now what had happened.

"Father," I called through the wall to my father who was inside the wash room, "come out here and look. These bees have all been poisoned!"

"You're right!" my father agreed as soon as he had come around the corner of the building and taken one quick look. "This is a mess!"

Just then the caretaker came to look.

"These bees have all been poisoned," I told him. Blank astonishment crossed his face.

"I didn't do it!" he said. And I knew he was telling the truth. "And I don't know of anyone else who poisoned them either," he added. "But two years ago we had a swarm of bees in that wall and we got a pest control man to eradicate them. And then last fall we had another pest control man come to treat for bugs. Do you think he could have sprayed poison into that hole?"

"Very likely," I agreed. "That's exactly what he'd do thinking to forestall the possibility of bees entering again this spring due to the poison being already in the wall. But this swarm of bees went in anyway—and now they're almost all dead, all except the fielders who were away when the main swarm entered the hole. Poor bees!"

"What are you going to do with them?" the caretaker asked.

"All we can do is dig a hole and bury the whole mess, dead bees, honeycomb, honey and all," my father remarked sadly.

The caretaker pointed out a place where my father could dig a hole and I got to work and cleaned out everything in the wall from the floor up into the high corner of the diagonal bracing. I have a special long flat-bladed wrecking tool that I always carry along for such an emergency as with it I can reach up or down into narrow confined places and thoroughly clean them of all bees and old combs. When I was finished I saw the owner's assistant coming toward me.

"I hear that you didn't get any bees," she said. "Too bad. We had surely hoped that you'd get them, both for your trouble and because we hate to see bees killed. Can you do the job for the twenty-five dollars agreed upon?"

"Yes," I told her. "And it's been nice working for you. I just wish, too, that we could have saved the bees. Can you get the wall all replaced again?"

"Yes," she answered, "Charlie will fix it. Thank you."

It was a paying proposition for the owner to give me the sum asked for I had now removed all of the dead bees and old comb that filled the wall and which, if not removed, would have been a constant source of trouble from ants who, for years to come, would have kept trying to get at the honey. Also, when the sun shone warmly against the side of the building, odors from the wax and honey inside would have wafted out through the cracks in the wall to alert other swarms of bees to the possibility of making the same wall their home. 🐝 It is always good policy after bees have been removed, by whatever means, to open the wall and clean it.

We loaded our equipment and empty beehive and went home. That afternoon my hand, due to the poison that had been sprayed upon the old combs, swelled up like a toy ballon. But it did not hurt me. Several people who saw it marveled that I had no pain. By noon the next day the swelling had gone down and I had no bad aftereffects, praise the Lord. Sometimes I have felt ill for as long as a week after getting into a poisonous situation. 🐝 It is due to the possibility of old poisonous compounds previously applied to the interiors of walls and attics that always prompts my father and me to open the bee infested area from the outside of the building, never from the inside. Homeowners sometimes correctly argue that it would be easier for them to replace a four-by-eight foot panel on the inside of the building rather than a section of siding on the outside. I agree with them in that aspect of the situation but it is the beekeeper who must finish the job of removing the bees once he has started it even though he discovers to his dismay that some part of the wall is contaminated with insect poison or repellent, and he himself may be very adversely affected thereby. One must make a charge for doing such bee removal jobs. At the present time the fee is usually somewhere between twenty-five and fifty dollars depending upon the accessibility of the bees and whether or not there is any real possibility of successfully hiving them.

Smoking Bees from Walls

Many people have asked me if I have ever really succeeded in using an ordinary bee smoker with burlap for fuel as a means of smoking bees out of a wall. Yes, I have done it and others have done it. 🐝 But there is a great risk involved, namely, fire. A few weeks ago I had an ideal situation. The bees were again located in the wall of an outbuilding close to one corner. They could not get up into the attic nor could they crawl down under the floor. If bees have a back way out to either of these places we can never smoke them out. The bees' entrance hole was about halfway up the wall. I quickly bored a one inch diameter hole high on the wall and another hole down low just above the floor level. Then I started up my smoker and really poured in the smoke both into the hole above and the hole below. Almost immediately the bees began to come out of their in-between entrance hole in great numbers. It looked like an easy hiving. But then I saw a wisp of smoke that had a darker bluish color come out of the upper hole—and burlap smoke is white. I jerked the smoker nozzle out of the lower hole and realized I had worked the bellows so rapidly that little tongues of flame must have spurted out and into the wall. After an anxious minute the blue smoke from the upper hole subsided— I had not set the house on fire—but I had come mighty close to doing it. Suffice to say I did no more smoking but sawed the wall boards and got at the bees that way. Also I wanted to be right sure that there was no smoldering fire left in the wall.

A much better and safer idea might be to use a product called "Bee Go" to chase the bees out of a wall. I recently bought a pint of this chemical and have loaned out small quantities of it to beekeeping friends of mine who do bee removal work. This chemical has a horrible smell which bees cannot tolerate. As soon as a swab or other cloth is saturated with this chemical and placed in the vicinity of bees they will move out—and fast. Diamond International Corporation, Apiary Department, P. O. Box 1070, Chico, California, 95927, carries this product. The advertisement in their 1977 catalog reads: Bee Go is a chemical used for chasing bees out of honey supers. Fast acting, keeping bees quiet, in all types of weather. Very easy to use. Fully approved USDA and Pure Food Administration. Cat. No. A-593, 1 pint Bee Go, Ship. Wt. 2 lbs. Write for price and shipping charges.

Removal with Bitter Almond Oil

Jim Meyer who does bee removal work for hire in this area just now told me of his latest experiment in removing bees from walls. Instead of using smoke, Bee Go, or opening the wall, he makes an inch auger hole in the wall and inserts a swab on a stick that has been saturated with bitter almond oil and the bees come right out ready to be hived. He says he really likes this method as the almond oil has a pleasing scent and may be procured from many larger health food stores.

Consider Your Bees' Welfare

Usually hiving a swarm of bees from inside the wall of a house takes all afternoon, and the job should be finished that same day. If one runs into unexpected problems and it is getting late toward evening the best thing to do is to help the bees back into their home in the wall, wait a few days for them to recuperate, and then try again. Bees cannot be penned out of their home overnight. They must be able to cover and warm their brood, as well as feed and protect themselves. They must not be exposed for hours to the direct rays of a hot sun or to the prolonged chill of night air. Otherwise too many of them will weaken and die before they can be hived. Always remember that bees are small creatures and quite fragile, so use a minimum of smoke, Bee Go, or roughness in getting them out. Their total life expectancy is short, at best.

When I remove bees from the wall of a house it always takes me at least four hours to do the job. I plan to begin at one o'clock in the afternoon, and if all goes well, finish by five o'clock. If I run into unexpected difficulties such as the bees being farther back in the house than the outside wall, it may well be seven or eight o'clock before I finish. If I see that the job cannot be completed before nightfall I stop for the night when all of the bees have some protection, either in my hive box or protected from the night air in their home in the wall, or a combination of both.

Honey in Walls

A beekeeping acquaintance stopped by to see me one beautiful June day. He had heard of a large old vacant house that was the home of several swarms of bees. As Roger wanted to increase the number of hives in his backyard apiary he had gone to look at the house and found the bees busily coming and going through several large cracks in the wooden walls. Since these various entries were some distance from each other and on different walls, he was sure that there were indeed at least three separate swarms living in the old house. He learned that the property had recently been sold and the new owner was planning to have the building bulldozed to the ground and burned. Roger found the new owner and received permission from him to remove the bees from the walls using any means he wished, even to tearing off part of the siding in any kind of a sloppy manner, as the house was to be wrecked anyway. But time was running out. A bulldozer operator had already been engaged to do the demolition work and would begin as soon as he could get his equipment on the job. So Roger came to see me for some of my free advice.

"I want to try to get a swarm of those bees and their honey," he told me. "One wall has a great number of bees coming and going. I just might get a lot of good honey, and a swarm too. I have an empty hive."

"Then by all means try to remove that large swarm first. You may really have found something."

"But what'll happen if I tear off the siding and find a lot more honeycombs than I'll need to tie into the empty frames of my hive as I hive the bees? What can I do with the rest of the gooey torn up combs and drippy honey covered bits of board and rubbish?"

"If you have to get the bees out in a matter of hours you may not have the time to do everything just as you'd like. But by all means cut out and tie into your empty frames as much of the sealed brood as you can—the more the better—and then scoop up bees with a pan or anything else you have handy and dump or brush them over the tied in brood to start the bees coming toward your hive. With good fortune you'll get the queen too, or she'll come of her own accord to the bees and brood in your hive box. 🐝 Then as fast as you can cut out the honeycombs from the wall whether they be good, bad, black and broken, or whatever, and after shaking and brushing the bees from the combs you've cut loose, onto and over the frames in your brood chamber, place them in buckets, a tub or dishpan, or whatever else you have available including cardboard boxes, and take them home."

"What'll I do with them after I get home? Won't my own bees smell them and be a humming horde in a matter of minutes?"

"Yes. Quickly transfer the containers of comb from your car to your back porch or even the kitchen where your own bees can't get at the honey until you're ready, and them remove the combs from your large containers placing them into other containers small enough to easily fit inside an empty super box. 🐝 Then on the next warm afternoon place an empty super box on your newly acquired hive. In this box place a container or two of your salvaged honey covered combs. Replace the cover and you're done. This is the easy way to give your newly acquired hive a real boost."

"I've a feeling I'll get a lot more honey than I can feed to my new hive."

"Good. In that happy eventuality give the balance of the combs to one of your other strong hives. Place an empty super with fully drawn combs on top of the hive you select to do the cleanup work, and then on top of that add a super box without frames into which you can place containers of salvaged combs until you have it all fed back to the bees and they in turn have it all stored away in their own extractable frames. 🐝 Bees can usually clean up a container of piled up wax fragments in two days so plan to remove the cover and add more every other day, removing the first container and replacing it with a second pre-filled container. That way you can get the exchange made with the least disruption to the bees."

"Man, I'm keen to give it a try!" Roger exclaimed.

"If you really do strike it rich you may have to add one or two more drawn-comb supers for the bees to use to store the honey."

"If I get that much honey I'll come to see you again!"

"Good enough! In any event let me know how you make out."

Some weeks later Roger told me that he had only had time to hive the largest swarm of which he had spoken. The bees had been gentle and he had hived them without difficulty. He found that the bees had stored a large quantity of honey in what proved to be an unusually large open space, part of which he would have thought should have been part of a closet. But apparently the original builder had other ideas. As a result Roger had been able to salvage a good ten gallons of excellent honey. He was delighted, of course. It had taken his bees almost two weeks to separate the good honey in his containers from the myriad bits and pieces of good honeycomb, old black comb containing some honey, honey covered splinters, bits of honey soaked dry-rot wood, and everything else that had become mixed up in his honeycomb containers as he hurriedly tore off the wall boards and salvaged what he could.

Bees In Walls

Roger was fortunate. To salvage that much honey from one swarm in a wall is rare in my experience. Most of the time if we can find one or two gallons we consider ourselves lucky. As a rule the space between the studs of a house is not large enough to allow the bees to store much excess honey for you and me. But once in awhile one strikes a bonanza and that is what keeps some beekeepers always on the lookout for bee infested houses, outbuildings, stumps, and bee trees.

Hiving bees from the wall of a house is one of the most varied and difficult of all the situations that confront a beekeeper. Anyone who has successfully completed such an undertaking is well on the way to becoming a knowledgeable apiarist. And you can do it too. Careful planning plus determination are the keys to success.

5

How To
Tame Honeybees

As TIME PASSES since the publication of our first book on beekeeping, more and more people from all around the United States and from other parts of the world as well, have come to see us to discuss our methods of handling honeybees. We are always delighted to meet these people but they do pose a continuing problem of how to keep our bees favorably inclined to their presence. Our bees must continually adjust to ever changing variations in the sound of the voices that come near their hives, variations in the scent of the breath, the clothing, and the perfume that our visitors exude, and differences in personal appearance. So last spring when an ever increasing number of strangers came to see us it was no wonder that some of our bees became concerned and buzzed around to try to ascertain the source of so many unfamiliar sights, sounds, and scents, especially in the immediate area of our back door. Our world record winning hive No. 4 of 1974 was only ten feet from our back door and one hive was even closer than that. These were the ones that showed the most concern about the presence of so many strangers.

"The bees are beginning to buzz our company rather closely," I remarked to my father one morning in May of 1976. "Do you think there is something we can do to quiet them even more than we're doing with our "wave cloths" hanging near the hives?"

"Yes," he answered, "and I think I know what it is. I'll go right out to the shop and make a more sophisticated wave cloth than any we've used to date. I've been giving this problem much quiet thought for the past two weeks."

In an hour he came back to the house carrying an eight foot length of three-quarter inch diameter iron water pipe, a post maul, and two thirty-six inch lengths of one-quarter inch diameter iron rod each of which was bent at a ninety degree angle for the last fourteen inches at one end. A loop was fashioned in each rod at its outermost end. In addition he had a wire clothes hanger, part of an old shirt, and half of a burlap gunny sack.

"What're you going to do with all that stuff?" I could not help but smile as I looked at him from the back door.

"Come along with me and you'll find out. Let's try my idea on some hives a little farther from the house."

We went to the end of one of our rows of hives. There my father placed the pipe in a vertical position.

"Climb on that chair, take the maul, and drive this pipe about two feet into the ground," he directed.

The ground still being somewhat wet and soft I soon had the pipe driven to the depth requested. Then I watched as my father lowered the bent ends of his iron rods into the top of the pipe, hung the clothes hanger with its old shirt in one loop, and on the other rod tied the piece of gunny sack. Then he tied a thin piece of board like a stake across the top of the pipe and to each of the rods in such a way that they would be held firmly end to end in a straight line. A gusty wind was blowing and in a moment his scarecrow-like contraption swung almost around—and then back again as another gust caught it from a different angle. It made a truly super wave cloth. Both of us laughed at its strange antics caused by the fitful wind. In a moment several bees came to investigate this new phenomenon in their immediate vicinity, but since it seemed to pose no harm they soon became accustomed to it and went on with their work.

We have used this latest wave cloth with complete success for the season just past. We can change shirts or sacks in a few moments with variations in color, scent, weight, and general appearance. As a result our bees pay little or no attention to people passing almost right in front of them with just a four-foot-high redwood shake fence located five to seven feet in front of the hives separating our bees from our visitors.

But I should also say right here that I do one other thing that, used in conjunction with the wave cloths, probably keeps our bees more gentle than they would otherwise be. When no one is here for a visit I often walk around inside the "bee pen" as one might call the enclosure surrounding our hives. I walk right in front of the hives among the swirling hundreds of bees as they leave or return loaded. I never wear any protective clothing whatever for if there is a cross bee among the thousands in my hives I want to know about it and I want that unhappy bee to sting *me* rather than one of our visitors. To make the test more realistic I wear all kinds of clothing from clean polyester to truly soiled woolens. Woolens still seem to anger our bees more than any other fabric. So I often wear a woolen shirt, and sometimes a whole woolen suit of clothes when walking among the bees to condition them to accept me. And they will, and do, but obviously grudgingly. Some days they buzz me in a really frightening manner—but so far none has stung me this year—not even one. I cannot help but say, "Bless their dear little hearts!" They know I am their keeper and they will not sting me even though they long to do so in order to rid an unpleasant scent from their immediate presence. Truly bees are kindhearted little creatures.

Working Within the Bee Enclosure

However, as beekeepers, we must always remember, especially if the day is warm and the bees are flying strongly, that we should do everything possible to keep from antagonizing our bees. One of the most effective things to do if we want to build or repair a fence or windbreak very near our beehives on a warm sunny day is to thoroughly smoke ourselves from head to toe both front and back. I fire up our four inch by ten inch bee smoker and when it is going well and the burlap fuel is really throwing out a lot of smoke at every puff of the bellows, I either smoke myself front and back and especially my hands and face, or I have my father do it for me. Then I thoroughly smoke my father in the same way and we get to work. Then we can literally do anything we wish while right in front of our hives without irritating our bees. If our job takes longer than six or eight minutes to complete we smoke each other again, and possibly, if it seems advisable, I give each nearby hive a puff at its entrance also. In this way we continue to live in peace and harmony with our bees.

Sometimes when we mention that we smoke ourselves more than we do our bees, a visitor will exclaim, "Then if one smoked cigarettes he should be able to handle bees with ease, and not even need a smoker and burlap."

"Not necessarily so," I answer, "and for a very good reason. True, some who smoke can handle bees as easily as I do but not for that reason because none in my family has ever smoked, and I strongly advise against smoking tobacco or anything else when working with bees and honey because it's likely to taint the honey—and no one wants tainted honey!"

"But why wouldn't smoking help? You'd think it would be better than burlap," some insist.

Using Smoke as a Veil

"No, because we smoke ourselves only when we're going to work with our bees and this changes our body scent

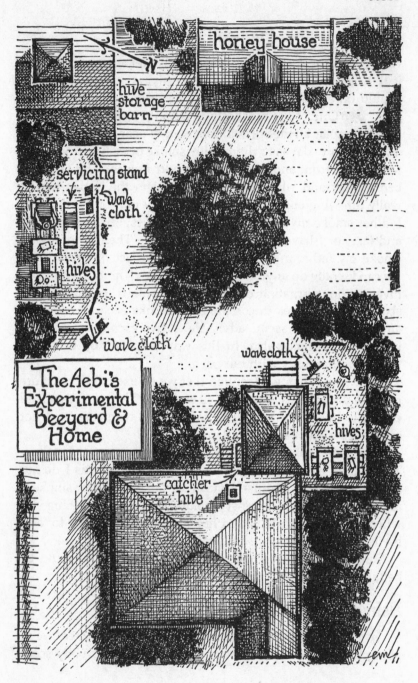

honey house

hive storage barn

N

servicing stand

wave cloth

hives

wave cloth

The Aebi's Experimental Beeyard & Home

wave cloths

hives

catcher hive

to that of a different person than we are when we daily walk among and talk to our bees. Bees consider anyone who breaks into their hive, for whatever reason, an enemy. So we change our odor and even our personality to some extent, at least it seems so to the bees, by smoking ourselves before and during our work time with them. Thus later they do not associate us as being the ones who disturbed them and they're not frightened or offended by our presence an hour later. The kernel of the matter is right here. If one smokes any of the tobacco products the bees soon associate that smoke odor with the man using it, and when he tries to take honey from his bees without using auxiliary smoke, such as burning burlap, they know who is the guilty party even though he's saturated with an odor of smoke. When he comes around again later the bees will remember him and he may well be stung as badly as though he had not smoked at all. Whereas when we use burlap to smoke ourselves up one side and down the other, our whole body odor is changed to that of the smoke used, and when we meet our bees later, our scent being changed, they do not associate our usual scent as being that of the one who opened their hive and annoyed them. As a result our bees continue to consider us as friends and we can come and go among them with our usual ease. But a continual smoker of tobacco or other such product has not changed his usual odor by smoking more of the same thing when he tries to work with his bees, and as a result, he can be—and often is—stung as though he had not used any smoke at all."

No Smoking: Bees and Honey Present

I have mentioned that those who smoke tobacco when they are working with their bees to take off supers may well inadvertently taint their honey. This is especially true for those who work or visit in the extractor room while extracting is being done. A few years ago my father and I were just finishing extracting a considerable quantity of honey and had numerous four and five gallon plastic buckets two-thirds filled with newly extracted honey on the table and shelves when I heard a car stop nearby. I had not had time, nor seen any real necessity, to cover the tops of any of the honey cans with either their plastic lids or white cloths as I thought I would do that when we were completely finished in another ten minutes. But just then someone called to us from the yard outside and I replied to him to come in. The man and several other members of his party came in to where we were working. He watched for a minute or two and then while my back was turned as I was uncapping a frame, he lighted a cigarette—one of the stinkinest burning tar pit El Ropos I have ever smelled. I instantly turned and asked him to put out his smoke but he just got red in the face and kept on smoking. With all of our delicious fresh honey standing all around I absolutely could not have him smoking in our extracting room so I quickly turned off my electric uncapping knife, took him by the arm and started him for the door, explaining as I did so that his type of smokes could ruin all of our honey in a very short period of time. Before he could really take offense I had him out of the door. Then I explained the foul odor taint problem more fully.

"Honey in open cans," I said, "or in uncovered plastic pails, or even honey in the comb, quickly absorbs unpleasant odors and if subjected to them long enough can, in our experience, so detrimentally alter the flavor of the honey as to make it ill-flavored and undesirable. As a result we are extremely careful in our handling of honey in all stages of its production. Honey extractor rooms have a naturally sweet scent that is almost heavenly—and we as honey producers must do everything possible to enhance and maintain that wonderful essence in our beautiful honey."

"I see what you mean," he said. "That's one of the reasons I drove all the way over here to see you. You always get along so well with your bees and have such good sweet honey too, and I never do. And I thought it was my bees that were at fault. Never suspected that it might be me!"

He and his group left as friends and I invited them to come to see us again.

There's One in Every Crowd

Occasionally we have a cross bee that our wave cloths will not quiet. This happens most often during the spring and summer when our hives are very strong. At such times we find a bee that continually buzzes us whenever we come near. We call such bees "little soreheads" because it is the same bee that threatens us time after time, and will even sometimes sting if given time enough. We find it necessary to swat and kill these few troublesome bees. This is easily done by using a small slat of wood one foot long, two inches wide, and one-eighth inch thick. A little piece of one-eighth inch hardwood plywood is excellent. When the bee comes buzzing threateningly close to our face we rapidly wave the slat back and forth in front of us like a fan. The bee is often attracted to the motion and flies into the path of the slat and receives a resounding swat. Some of these little soreheads pick themselves up and fly away to go back to their work. Others come at us again and we give them a harder blow the second time that ends their career. No beekeeper can afford to have a few disgruntled bees keep all visitors uneasy.

Using Machinery Near Beehives

Bees do not appreciate the noise and vibration of a gasoline powered rotary lawn mower near their hives. I mow the area near our beehives early in the morning just after daybreak. But if one must do the work later in the day it is always well to begin mowing in the far corners of the bee yard and work toward the bees by degrees with the mower speed set as slow as the engine will pull the load. I never use a veil or protective gear of any kind but it would be well for most people to do so. To obtain the most honey, grass should be kept mowed short at all times, otherwise the early morning field bees returning to the hive too often miss the landing board and get tangled in the wet grass below the hive and waste considerable time and effort extricating themselves. In many places the morning nectar flow is far greater than that in the afternoon so we want to help our bees all we possibly can to collect the maximum morning flow.

Bees Appreciate Cleanliness

Always wash your bee suit, if you wear one, or your clothing after working with bees to eliminate the odor of bee venom. Sometimes a few frightened bees will make defensive passes at their keeper and touch his clothing just enough to exude a small amount of sting fluid. If allowed to accumulate over a period of weeks this odor may become strong enough to really antagonize a hive of bees and cause them to come at us almost without warning, and without visible provocation. Hives vary greatly in their sensitivity to this old venom odor. If one gets a "tick-type" sting, that is, one that penetrates the skin but does not cause the bee to lose its stinger, always quickly wash the area with water so as to eliminate the odor of the poison left by the bee. Otherwise other bees flying nearby will scent the poison odor and will be inclined to give us or our clothing other tick-type stings until the odor becomes strong enough around us to make the bees really want to sting in earnest. A quick puff from the smoker will also do much to hide the odor of a tick-type sting. I never wear a bee suit, so usually get such stings on my shirt. As soon as possible after finishing with one hive, even if I still have another hive I want to examine, I go to the water faucet and wash down that spot on my shirt to eliminate the odor. This is a great aid in keeping our bees gentle. If one has many hives and is working away from home in an outyard far from neighbors this precaution is not so important. But the backyard beekeeper must keep all of his bees gentle.

Disturbing the Peace

Bees resent any sudden blow or jar given to their hive. Recently a friend of mine wanted to move his hive of bees about four feet. The hive was on a wooden pallet and he found it too heavy to move by lifting alone. So he used a piece of board and nudged and jacked it over a few inches at a time. This series of short jarring lifts and drops really infuriated his bees and in spite of his bee suit they managed to find ways to sting him and they did so severely. He had no really bad reaction from almost a dozen stings but he had learned a painful lession.

Taming Truly Wild Bees

On October 5 my friend Bob Jett came to see me. By the strained look on his face I knew he had a problem.

"What's the trouble, Bob?" I asked him.

"It's those wild bees I bought and brought home three days ago. They've kept all of us penned up for three days, and all of us have been stung—my wife, three children, the dog, the cat—and even the chickens. And if any of us venture outside we get chased right back inside again or get stung!"

"Where in the world did you get wild bees like those?"

"Up in the hills away back of my place. The owner didn't want them and offered them to me at a very reasonable price so I bought them, thinking to place them beside the hive I already own and so have a chance to watch two hives at work this winter and spring. But I'd no idea these bees would be so vicious and sting everything in sight."

"You really did get a wild swarm," I agreed.

"What can I do now? We're beginning to get desperate. My own bees are so gentle we can all watch them and I had a comfortable chair placed only ten feet away from the hive and could watch them all day or any time of the day and never even thought that they might sting for they never did. But now—look out!"

"What you should probably do is requeen them right away with a more gentle strain," I said, "but the trouble is I don't know of anyone who has a queen to sell this late in the season, and I don't have one either. The only other thing I can think of is for you to tame your wild bees but that will take from one to three months."

"Can you tame *bees*?" Bob asked in amazement.

🐝 "Yes, but it does take time. This evening after the bees are all in their hive for the night, go out and drive eight six-foot-long stakes into the ground in a semicircle about five feet in front of your wild hive. On these stakes hang a soiled garment, the more soiled the better, from each member of your family including the dog's sleeping rug, the cat's scratch post, and something from the chicken house."

"Do you mean that I could even hang some of our *underwear* close in front of the hive?"

"Yes. Soiled underwear is excellent. But a sweaty shirt, pants, or whatever else your wife will let you hang out for a month will do."

Bob laughed. "May I ask why?"

"Yes. It's so that the bees can get a close-up view of human wearing apparel as well as a strong scent of the odors that so much frighten and offend them. When the sun comes up warmly tomorrow morning the bees will probably come out of their hive in a fury and sting every item of clothing you've hung up. But since neither you, nor your children, nor any other member of your household will fight back at the bees, or run away, or give any indication of trying to harm their home, the bees, being intelligent insects, will begin to realize that they're needlessly fearful and will begin to accept your second-hand presence and in time will accept you and all in your family."

"I'll go right home and try it!" Bob cried in relief and eager hope. "I do hope it works!"

Two weeks later Bob came to see me again and this time he was in a jubilant mood.

"It worked!" he exclaimed. "It really did! It took just seven days for the bees to accept all of us. Now I can again sit in my easy chair and watch both hives and my dog lies at my feet and watches them too. No more problems whatsoever."

"That's indeed good news," I agreed. "And I'm so glad you came to tell me. But how do you account for your success in quieting your bees so quickly?"

"I have a family, and you don't. Could that be the reason?"

"Probably! Bachelors don't have wives and children so I never had enough garments to hang up to try the experiment to its fullest extent. Congratulations, Bob, on taming your bees!"

Artificial Scents

Another day Bess Luker stopped by to see me. She has had bees for some years and when passing my house sometimes drives in to see how my bees are doing, and I also have a chance to check on hers.

"I see your bees are hard at work," she said as she stepped from her car. "Mine are working too and I sometimes wonder what they can find at this time of the year—" Before she could say anything more I saw a bee in the act of passing her, swerve and sting her on the neck.

I jumped toward her and instantly scratched out the stinger, at the same time exclaiming, "I'm sorry! I wonder why one of my bees did that?"

"Think nothing of it," she said, "if I were at home just now my bees would sting me too."

"But why?" I cried.

"Because of this particular brand of perfume I'm wearing. I like it, and I wear it when I'm away from home, but my bees dislike it intensely—and I see yours do too."

"Well this is the first time that I've ever heard of perfume offending bees," I said.

"I know," she replied, "but I can expect to get stung when I wear this perfume. I had not thought of stopping to see you when I left home this morning. But don't you think it has a really exquisite fragrance?"

I stepped a little closer to her and drew a deep breath. "Yes it does," I agreed. "I think it's wonderful!"

"But bees never do! So I'll cut short my visit for this time and see you another day."

Life Cycles

In July of 1976 I attended our Advent Christian Campmeetings that were held here in Santa Cruz, California. One of the Bible teachers was Pastor Ransom from the Advent Christian church of Napa, California, and he spoke one morning on the subject of biorhythm, a subject of much successful experimentation in Japan. The point of the study is that on certain days, at predictable intervals, we, as human beings, are prone to be cross and irritable. And if a man and his wife are rhythmed so as to be cross on the same day—look out! This really startled me, for I have observed the same thing among my honeybees. Some days, most in fact, I can go among them with perfect ease and they seem glad of my presence. But there are days when they become cross as I approach. So I leave them alone that day. If I had intended to open a certain hive and take off honey or rearrange the supers, I put it off until the next day at which time the bees are usually happy again and I can work them without protective gear. This policy also keeps them from becoming unhappy with our many visitors.

I do not yet know the time interval in days between our bees' cross periods, but I am surely watching and am making a real effort to find out. I would urge everyone who reads this account to also observe his bees on this point. It would be of tremendous help especially in highly populated areas if we knew in advance on which day our bees would be the most gentle. It would save many a good neighbor from becoming unhappy with us if we could pick a day on which he was happy and the hive we want to work was also happy.

I remember so well a friend of mine who came over to see me, and immediately said, "Boy, did I get a stinging today! Oh, I had on all my protective gear and didn't actually get stung so it hurt but my bees were mad at me from the beginning even before I began to remove the hive cover or supers."

"Surely you didn't work with them when you recognized they were in a bad mood, did you?"

"Yes I did! I'm just as stubborn as they are. I figured that if they wanted to sting to go at it as they couldn't hurt me, and they didn't—" he paused.

"And then what happened?"

"They stung my mom between the eyes when she went out for the mail an hour later, and they buzzed my dad like crazy when he came home in the evening, and even now when I left home to come to see you they would have stung me if I hadn't gotten out of the driveway fast!"

"Yes," I replied gravely, "and they'll be angry for the next three days or I miss my guess." And they were indeed angry for the next three days as he told me later.

Until Pastor Ransom spoke of biorhythm I had never heard of it but I am sure that what he said about human beings is also surely true of honeybees. 🐝 For many years, before working with a hive, I have gone an hour earlier and blown my breath near or into the entrance of the hive to test the bees for love or anger, and I have never yet received what beekeepers call "A stinging," that is, two or three dozen stings within a matter of ten or fifteen minutes. And I never expect to receive such a stinging especially now that I know a little about biorhythm. Let us all endeavor to work our bees on their happy days.

Love Those Honeybees

Honeybee Quiz No. 1

TRUE OR FALSE. Now sit back and relax for a moment. There are several ways to increase your knowledge of the marvelous world of the honeybee. One way is to keep several hives of your own and spend a lifetime observing them. Another is to read books and articles. A third way is right here in front of your eyes—the honeybee quiz. Few of the questions have been covered thus far in the book so unless you are a student of honeybees you will probably not get all of the answers right. If you answer half of them correctly you will know more than the average person. A score of 15 is excellent. *Answers follow the quiz.*

1. All honeybees have stings   True_____   False_____.

2. Bees need a little salt to season their food.
True_____   False_____.

3. Honeybees have eight legs.   True_____   False_____.

4. There are many species of bees but only a few are known as hive bees.   True_____   False_____.

5. Bees cannot climb the glass of a windowpane.
True_____   False_____.

6. There are no bees in the state of Hawaii.
True_____   False_____.

7. Beeswax has a higher melting point than paraffin.
True_____   False_____.

8. Drones are the specialized workers who build the queen cells.
True_____   False_____.

9. Rape is a valuable nectar producing field crop.
True_____   False_____.

10. All honeys have the same color.   True_____   False_____.

11. Blue Gum eucalyptus trees, which produce much nectar, have pale yellow blossoms.   True_____   False_____.

12. Some species of bees live in the ground.
True_____   False_____.

13. Many people prefer crystallized honey to liquid honey.
True_____   False_____.

14. Some oak trees produce pollen.   True_____   False_____.

15. Honeybees cannot live at an altitude of more than 3000 feet.
True_____   False_____.

16. Black plastic sheeting wrapped around a beehive will absorb heat from the sun.   True_____   False_____.

17. Bees obtain pollen from the blossoms of wild blackberries.
True_____   False_____.

18. Desert sage produces an excellent honey.
True_____   False_____.

19. Queen bees live only one year.   True_____   False_____.

20. A strong hive of bees would number about 5000.
True_____   False_____.

How To Tame Honeybees

Answers:

1. False. Male bees, the drones, have no stingers. And visitors tell me there is a species of honeybee in South America that bites, but has no sting.
2. True. A dash of salt on the landing board every four weeks for maximum health and honey production.
3. False. Bees have six legs.
4. True. Most bees and their relatives the wasps are solitary or very small colony insects.
5. False. Bees can crawl up, down, or sideways on glass panes.
6. False. Hawaii produces a commercial honey crop.
7. True. Paraffin melts at about 110 degrees, wax at about 150.
8. False. Drones do no work in the hive.
9. True. Rape (*Brassica napus*) of the mustard family blooms at four weeks and grows and blooms sometimes for months.
10. False. Honey colors vary greatly. Also there is a great difference in flavor.
11. True. Usually about two inches in diameter with much nectar.
12. True. As far as I know these are always solitary bees.
13. True. We sell more than half our production as crystallized.
14. True. Some years our California live oaks produce much pollen.
15. False. A friend in Leadville, Colorado, elevation 10,000 feet wrote me that there are wild honeybees near his home.
16. True. We sometimes find black plastic very useful in bringing a weak hive through the winter.
17. True. But the pollen is so dark gray that one may not see it.
18. True. We have seen our bees work on white, black, and purple sage.
19. False. Queens may live three or even four years.
20. False. A strong hive would number 80,000 and up.

6

The Sanctity Of The Hive

Working a Large Hive Midsummer

ONE WARM JULY AFTERNOON three men stopped by to see me work with one of our huge hives that still had nine medium depth supers above the queen excluder. They were desirous to see how I would work a hive that, on its stand, towered more than two feet above my head. I told them to stand near to one side close to where I was going to work. 🐝 I then arranged two five-foot stepladders four feet apart with their steps facing each other. On the third step up I laid a one and one-half inch thick by twelve inch wide plank, and after thoroughly smoking myself from head to toe I climbed up onto the plank and prepared to remove the cover from the hive. But before actually removing the cover I arranged two V-shaped one-by-one-by-eighteen inch pieces of board across the top of each stepladder so that when I had a super removed I could place it upon the strips of wood and so leave a space beneath the super thereby avoiding the almost certain risk of crushing some bees when I set the heavy super on the flat ladder top. We always make special enlarged ladder tops capable of holding the weight of several supers piled one upon the other, each being separated from the one below by our V-shaped strips of board. 🐝 Thus we do not have to climb down each time we remove a super to place it on the ground, and then minutes later, carry it back up to its place on the hive again. Also to further facilitate removal and replacement of supers my father built me a lightweight wooden super holding stand four feet high, twenty inches wide, and five feet long. This stand is well braced with one-by-three inch boards and has three one-by-four inch boards spaced on edge equidistant lengthwise of the stand on which to rest the supers when I remove them from the hive. Thus I can quickly swing a super off a hive and have a handy place to set it down until I have finished servicing the hive.

The men watched me in silence as I lit the smoker, smoked myself thoroughly and then smoked the entrance of all the hives within twelve feet before proceeding with the working of this huge hive. I proceeded to remove all nine of the supers until I was down to the queen excluder where I swiftly swung a recently extracted super into place and then quickly replaced eight of the supers I had just taken off. 🐝 In taking off the original nine supers I had been careful to note which ones were partially filled, well filled, or completely full of honey and had segregated them into different places, the full supers being placed on the ladder top to my left, the well filled ones to the ladder top on my right, and the more or less empty supers I placed on the super stand behind me. Now as I began to replace them I placed the completely empty extracted super just above the queen excluder as already stated and then added the next least filled super and so on in order of their weight until I had the heaviest and most nearly filled ones at the top of the hive. One very heavy and well filled super that had originally been near the top of the hive was ready to take to the extractor room and extract so I left that one off and the hive when rebuilt again had nine supers. All this rearranging had taken me approximately ten minutes. I had used a little smoke as needed and

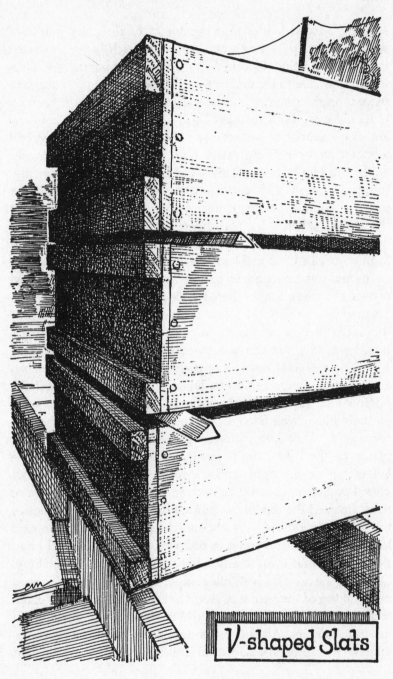

Aebi

V-shaped Slats

also my bee brush and had crushed very few if any bees.

"That was a quick job!" one of my visitors exclaimed. "But why didn't you go on down and look in the brood chamber? You were all the way down there and could have checked out the queen, eggs and brood in only a few minutes more."

"Of course I *could* have," I replied, "but why would I want to do that? 🐝 I already know the queen has been laying thousands of eggs each day because there are so many thousands of bees coming and going from the entrance very few minutes. And all of the supers have some honey in them. Four are heavy, and the top three are almost ready to come off. I checked the supers when I took them off and again when I replaced them. To check more than that is a useless disruption to the bees. My father and I never check into a brood chamber when the bees are active and strong, for there is no need. We have long ago learned to curb our idle curiosity." And I gave them an inquiring glance.

The man who had queried me about not checking the brood chamber smiled a little ruefully and turned a bit red in the face. I knew I had hit a sore spot in his experience and I wondered if he would tell me his story. After a sidelong glance at his companions, he did.

Disastrous Brood Chamber Meddling

"You know," he said, "I have three hives myself. All of them came through the winter in good shape but one was even stronger than the others. One clear day in the forepart of March I decided to open that strong hive and see what made it stronger than the others. I took my time and removed all of the frames in the brood chamber one by one and examined the combs until I found the queen. I also found some sealed brood, many larvae, and quite a few eggs. I put the hive all back together, but you know, that hive seemed to have lost its pep. A week later I went all through it again frame by frame and again saw the queen so she was all right but the hive still continued to decline in activity. About two weeks after that I went through the hive for a third time but didn't see the queen. Now it's the weakest hive I have and I think I may be going to lose it altogether."

"Do you expect to get any honey from it at all this season?" I asked.

"No. None whatever. It's just barely alive right now. Do you think I did something wrong?" he asked as he saw me shake my head in dismay.

"Yes, I'm afraid you did—possibly several things. First of all there was no real need for you to tear into your strong, obviously prospering, hive. 🐝 My father and I never open our very best hive just for the sake of curiosity, or even observation or experimentation. We always open our second best or third best hive, or even farther down the line than that. Was there a wind blowing the first time you opened your hive?"

"Yes. Not a hard wind but still somewhat cool, but the sun was shining."

I considered his answer. "How long did you have the hive open?"

"Oh, about twenty minutes, I'd say," he answered.

I nodded. "Probably the wind was chill enough to kill all of the eggs and larvae and even some of the sealed brood. The forepart of March is early in the season even for this climate and the air often has more of a nip to it than we realize. You may even have chilled the queen on your second or third examination thus causing her to be unable to lay many, if any, eggs."

"But I don't think it was that cold those days," he protested.

"Well maybe not. You could have done something else wrong that's often equally fatal to a strong hive."

"What?"

Keying the Brood Frames

🐝 "Failed to replace the frames back in the hive in exactly the same order and position as they were when you removed them. Were you careful on that point?"

Servicing Stand

"Never gave it a thought. I put them back any way it came handy. Why be so careful about replacing the frames?"

"Because when the bees originally drew out the starter sheets they were probably not able to draw the combs completely straight as they should have done, and would like to have done. As a result they drew out the combs as best they could in as favorable a manner as possible to facilitate maximum brood rearing, each comb being spaced with great exactitude in relation to every other comb. When we remove the bees' meticulously drawn combs and replace them hit or miss we generally miss—we fail to replace the combs so that the spacing is exactly as it was originally. Before hiving a new swarm my father and I always draw a heavy black line diagonally across the tops of the frames in the brood chamber so that if we have to remove them later we can tell at a glance if we're replacing them in order as we should be doing, or if we've misplaced a frame."

"Why is that so important?"

"Because, first of all, we usually have to crowd the frames close together to get them back into the brood chamber. And if we don't exercise great care we'll have the wax combs of some frames so close together as to be touching or even crushing against each other, and other frames so far apart as to make it almost impossible for the bees to cover their eggs, larvae, and sealed brood. When the bees have an ideal situation they space their drawn combs so that two bees, back to back, completely fill the space between the frames thus requiring a minimum of heat and effort to keep their brood warm and healthy.

"In the second place, we too often risk pressing or crushing the queen between the bees and combs or the combs themselves as we replace them. An injured queen may live for some time— long enough for the available eggs to hatch and the larvae to become overage for queen rearing. Then when the bees do realize that they're going to have to replace their queen they find themselves unable to do so. I think that's what may have happened to the queen of your strong hive."

"I think so too," my visitor agreed sadly. "I've suspected for some time that my hive might be queenless. But what puzzles me is that when I opened the hive for the third time I saw about a dozen open queen cells and some of them looked newly built."

"Yes," I replied. "We often find such empty queen cells, especially in a situation such as you've described. But if you'd looked closely I think you'd have seen that none of them had ever been capped or occupied by a queen because the old queen was no longer able to lay an egg in even one cell."

"Say I think you're right!" my visitor exclaimed. "I noticed that none of the cells seemed to have actually been occupied. And you think I've been the cause of all the trouble?"

"I'm afraid so," I had to admit, "but take heart, we've probably all done it at one time or another. Your equipment is still good and useable and probably you can catch a swarm from one of your other hives after they build up strongly."

"Thank you for the encouragement," he said as he looked at his friends. One of them had been looking at me rather steadfastly for some minutes and I knew he had a question.

"What is it?" I asked.

"Well you said awhile ago that bees draw out the comb in the brood chamber very carefully and we should replace frames with great care. But a few days ago I looked in one of my hives and was surprised to find the combs in what I thought was really bad shape. There were even areas in the frames where the comb was missing altogether."

"Right. You were looking in a hive that had been occupied by bees for several years or longer. As the years pass the bees often cut out and remove part of the old wax in the brood chamber. I think they plan on replacing it, as they sometimes do, but at other times something seems to interfere and they never get the new wax drawn. That's why my father and I always replace the old brood chamber with a new one between the fifth and sixth years for maximum honey production."

"Thank you for your time and information," they all said as

they bid me goodbye. "We'll hope to see you again."

Let me say right here that before opening a hive we should always consider carefully whether it is really necessary to open the hive at all, and what effect opening the hive will have upon our bees. Sometimes we must open a hive, of course, to add or take off a super but rarely do we need to dig down into the brood chamber except for the purpose of studying our bees or to note the condition of the brood frames. My father and I never risk our honey production for a whole season by opening and needlessly examining our best hives. After all, honey is money, and we have many eager customers coming, all of whom want at least a little of our good honey. And as a bee-keeper you will have such customers too.

Full Depth Brood Chamber Renewal

Every five years, as just stated, my father and I change brood chambers replacing the old black and rubbery combs with new fully drawn combs from the season just passed. To explain how we do this easily and quickly let me begin by an explanation of how we hive our swarms. We let our bees swarm naturally and when they have clustered somewhere in our yard such as on a trellised berry vine or fruit tree limb we place a full depth hive body with all ten frames in place just beneath the swarm. We remove the hive cover and shake and brush some of the bees onto the top of the uncovered frames. Soon the bees sink down between the frames and cling to the vertically wired starter foundation sheets. As soon as practicable, usually within ten minutes, we cautiously begin to slip the hive cover back onto the hive body beginning on one side and slowly edging it across to the other side. Then we brush some bees onto the landing entrance and they begin fanning to call all of the remaining bees flying around to come to their newly acquired home. Soon we are ready to stop off the entrance with a full depth and width wire screen closure cleat and take our bees home.

This single depth standard hive brood chamber will now be

the home of the queen and nurse bees for three seasons or three years, roughly speaking. During these years when the hive needs more room we add a queen excluder and supers above it as needed. ᗒᗕ We no longer use the standard double depth brood chamber for the good reason that we obtain a far greater maximum honey production with a single brood chamber. For those who are interested in the logic and theory of our method I will endeavor to expand upon it in a later chapter.

ᗒᗕ At the beginning of the fourth season the brood comb in the brood chamber is beginning to become quite black and hardened so we add a medium depth super of beautifully drawn combs just above it after having first removed the queen excluder. This enlarges the brood chamber and the queen will almost invariably move up into the nice bright combs to lay eggs sometime during the brood rearing season and once up there she will usually stay. Again we add supers above the queen excluder as needed, and remove them as the summer passes and fall approaches until once more we have all of the storage supers removed and our hive is down to its winter size of one full depth hive body and one medium depth super above it comprising the brood chamber. As a rule in our area and anywhere up and down the West Coast where we have lived— and we have moved fifteen times in the last fifty years—bees will store enough honey in this amount of space to provide them with food for the winter but if not, we leave the queen excluder on over winter with a fairly well filled medium depth super above it for additional stores. For the fifth season we again work the hive as we did during the fourth season.

ᗒᗕ Now at the end of the fifth season we begin to make preparations to shift the bees out of their old brood chamber quarters into new living space. We carefully examine our extracted medium depth frames and pick out those that are almost perfectly drawn and place ten of them into a medium depth super making up as many of these supers as we will need to replace the brood chambers of the hives needing a change. ᗒᗕ Then during the first weeks of spring buildup, usually sometime in February in our Santa Cruz area, we add one of our prepared supers. We lift the upper medium depth part of the brood chamber away from the lower full depth part and place our prepared medium depth super between the two, replacing the upper brood super as quickly as possible to conserve heat in the hive. Our hive brood chamber now consists of a full depth hive body and two medium depth supers. Above that we have the queen excluder and storage super that may have been left on all winter. ⚜ In a few weeks the queen will have moved up or down into the newly added combs. So sometime during the last days of February or the first days of March on a warm calm afternoon we take the hive apart setting the two medium depth super brood chamber compartments aside as a unit until we can detach the old hive body from its bottom board. ⚜ When this is accomplished we quickly and carefully lift the new medium depth compartments onto the hive bottom, then place the old full depth brood chamber immediately above them and smoke the bees down into the new area below. Usually the bees will quickly go down even though they may have some brood in the old chamber because the queen likes to lay eggs in new combs. ⚜ Then we add a queen excluder and replace the old brood chamber above it and also the winter storage super above that if the hive had one. This exchange of hive parts can be accomplished quickly, elapsed time often being less than ten minutes.

⚜ In ten to fourteen days, weather permitting, we open the hive and carefully examine the old brood chamber to make sure that the queen is down below the queen excluder and in her new brood chamber. To facilitate the examination we momentarily lift off the old brood box so that we can temporarily remove the queen excluder. This done we replace the old brood chamber on the hive again and begin to smoke the frames beginning on one side and working toward the other side. The smoke quickly drives almost all of the bees down into the new brood chamber below and if the queen is still among them she will also go down. Then we remove the frames one by one and

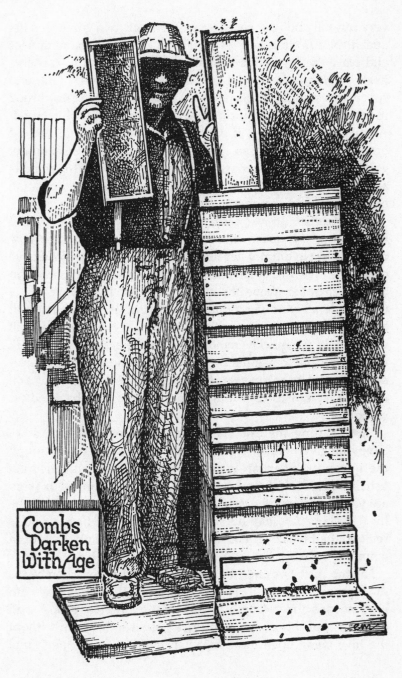

Combs
Darken
With Age

very carefully but quickly examine each one for a glimpse of the queen. Not finding her, we brush off the few remaining bees and place the frame in a suitable cardboard carton to conserve heat. We do the same with the balance of the frames. Care and speed are essential at this point for there is usually sealed brood in some of the frames and we do not want to chill this brood. At this time we rarely, if ever, find eggs or larvae in the old brood chamber. To do so would mean that the queen had not gone down earlier after we had made the original exchange, and is the reason why we make this final careful examination of the old brood chamber.

🐝 *Caution:* In rare cases we find that we did not succeed in inducing the queen to take up housekeeping in the new combs, and we did not get her smoked down either. When we check the hive weeks later we find her busily rearing brood in the old chamber above the queen excluder. If the hive is yet only moderately strong in numbers we find the queen and put her down below the queen excluder. But if the hive has built up strongly we quickly remove the queen excluder and work the hive for the remainder of the season without a queen excluder as we did for many years before excluders came into general use. We try not to discourage a prosperous hive by needless meddling.

🐝 During the examination of the individual frames we have also given the emerged drones, too large to pass through the queen excluder, an opportunity to escape—and they are usually quick to take advantage of their opportunity. If there are yet many capped drone cells we must continue to open the hive at about five day intervals. Usually all we have to do is remove the hive cover and the drones will come up and out in a minute or two.

For the next two months we use the old brood chamber as the topmost honey storage super slipping other supers beneath it as needed for the bees to use to store nectar, pollen, and honey. All of this exchange of hive parts can be done without the hive being open more than ten minutes at any one time, causing a minimum of disturbance to the bees. Actually we often accomplish our purpose before the bees are really aware of what we are doing. With our method we never remove any frames from any of the brood or storage chambers nor even look for the queen until the final inspection of the old brood chamber.

Medium Depth Brood Chamber Renewal

When another five years have gone by the exchange is even more easily accomplished. All we need to do is place two newly drawn supers between the two parts of the old brood chamber and after a few weeks make the exchange removing the old medium depth super from its bottom board and setting the three upper supers down on the bottom board and then adding the old super brood chamber as the fourth or top super below the queen excluder. In six days we remove the cover and smoke the bees down into the two new lower medium depth supers and the exchange is complete. There may be variations in procedure this time too, of course, but nothing difficult.

Salvaging Old Brood Chambers

Let me say again that in our experience it always pays to renew the brood chamber every five years. We find great variation in the condition of such brood chambers. Some would still be usable for another year or two, whereas the comb in others is to a large extent black and useless or even nonexistent, as the bees have already cut it out and carried it away and we find only empty frames or partially filled frames where fully drawn frames should have been, and had been, the year before.

So to gain the maximum honey production we arbitrarily change brood chambers every five years. In that length of time the bees will have used the same cells many times, and after each hatch of brood the bees clean and revarnish each cell making its walls ever thicker and darker with a slight lessening of available space for the larva to develop until the queen no longer can lay an egg in the cell, or at least she does not. Also during the years the bees store extra pollen not immediately needed in some of the cells, thus making them unavailable for egg laying and the number of usable cells often diminishes as the years pass. Often we see old and sometimes moldy whitish pollen stored in old brood frames. This pollen is totally unfit for the bees to use as food for their brood and it is difficult for them to remove from the hive. 🐝 Again we see very hard crystallized honey in some of the cells and this too further reduces the number of cells available for egg laying. But this honey, though stored in black unsavory looking cells, is excellent in quality and flavor and we always save as much of it as possible during our comb cutting out operation. 🐝 We drop all pollen-free chunks of crystallized honey into a container where we can crush them up at a later time and let the honey drain out through a large mesh colander. We cover our colander with a piece of clear plastic held tightly in place with a large rubber band and set it out in the sun to warm up. If the sun is shining brightly the crystallized honey soon melts and seeps down through the colander into the container below. On a more cloudy day I focus the sun's reflection from a large mirror onto the honey and wax and soon both melt and flow down. 🐝 After pouring off the honey I place my container of wax and residue in the oven to reheat to a higher temperature, about 300 degrees, until the wax melts and settles to the bottom. Heating wax in the oven takes careful watching and handling— I do not want to burn my house down. Solar heating presents no problems.

Wax in the Oven

On dark days I melt the crystallized honey and wax in the oven. 🐝 To keep the first liquified honey from overheating while the remainder is still melting I place in the oven a few inches above the burner a wooden board three-quarters inch thick by ten or twelve inches wide under my container of melting wax and honey. This board serves as an efficient insulating material to aid in keeping the already melted honey as cool as

Simple Solar Wax Melter

possible for the remainder of the melting period which may be as long as an hour if the colander has a gallon size or larger mass of wax in it. A sustained oven temperature of 300 degrees will brown the surfaces of a board, but after long use I have never yet had one take fire, due in large part to the fact that the oven is kept at a relatively low heat. But I do not go off walking or talking or hang too long on the telephone. One must remember his beautiful wax and honey!

After the honey and wax have both melted and seeped down through the colander into the container below I remove it from the oven. Because the pot is very hot to the touch I use leather gloves, or potholders plus additional crushed newspaper to protect my hands. I set the pot on a board on the sink worktable and remove the colander, placing it over a large disposable pie pan for the reason that drops of wax and residue continue to drip from the colander for some minutes until it cools and the contents harden, and one does not want any drops of this gooey material to fall upon the worktable or floor. Then for the next two minutes as the wax is cooling and the residue settling I quickly set out little plastic forms capable of holding about five ounces of melted wax. The plastic forms I like best are those used to pack margarine and other similar products. Such containers are made of flexible plastic and when the wax cools it pulls in the sides of the containers during the time the wax is cooling and shrinking. Later when I want to remove a wax cake all I have to do is press around the sides of the form with my hands and the wax pops loose.

After the two minute waiting period has elapsed I again pick up the pot of melted wax and hold it in my left hand at an angle so that the liquid wax is on the verge of pouring out over the lip of the pot into the little plastic forms mentioned above. I hold the pot for another fifteen seconds during which time the unwanted residue settles toward the bottom. Then I pour a thin stream of hot wax into the forms, using a spoon or wooden paddle to push back any residue that might try to float over the edge. When all of the good wax has been poured into

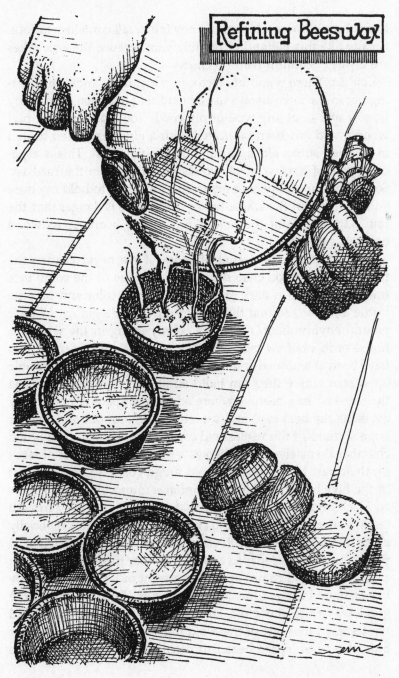

Refining Beeswax

forms I pour the residue and honey into a tall container to cool. I really like this method of refining wax because the wax cakes are always beautiful with no cracks.

Caution: There is another factor to remember. Under the melting pot in the oven always use a board that has been cut directly from a tree—not any type of plywood. Too many times a plywood board has been fabricated with a glue that when heated in an oven stinks like a burning or decaying fish. This is especially true of hardwood plywoods imported from the Far East. Sometimes upon opening my oven I have had to hold my nose and then air out the whole house. Also there is danger that the foul odor will taint the honey. But a pine board or redwood board is excellent.

After an extended period of years, if we never change the frames in the brood chamber, whole sections of the comb in a frame usually turn dark grayish or black in color and become brittle and hard so that the bees abandon it altogether. In their natural environment, such as a colony living in the wall of a house or in a hollow tree, we often see large areas of comb that have been abandoned and the bees pass over it to a yet available open area where they can build additional new comb to meet their needs. In a beehive where additional space is not readily available the bees can, and sometimes do, cut out and replace large sections of this useless old comb, thus renewing the brood chamber themselves. But it takes a prodigious amount of labor on their part to tear out such old tough comb and carry it out of the hive bit by bit. I have sometimes found old brood combs so black and thick and tough that it took all of my strength using a sharp butcher knife to cut it out of the frames, especially along the back bar of each frame. It is a question whether it pays a hobby beekeeper having only two or three hives of bees, to keep these old combs after they are cut out. Only a company like Diamond International Corporation of Chico can extract the good wax from these old combs. Once in awhile on a cold winter day I toss one of these old frames, wax and all, into our little sheet iron space heater. It catches fire quickly and burns furiously, thereby producing almost enough heat to compensate for the wax I waste.

If bees are left to their own devices to renew the comb in their hive they usually do it during the spring buildup. It is my observation that they do not enjoy this vast tearing out and rebuilding labor but would far rather attend to their normal activities of gathering nectar and pollen. So we help them with the hardest part of their work and in turn they use their time and strength gathering excess honey for us, their keepers.

Survival of the Fittest

Sometimes as winter passes into spring we find that some of our hives are very strong with much brood emerging, and others are as yet more or less inactive with few bees in evidence. My father and I do not rob frames of brood from strong hives and give them to weak hives. Never. Every one of our hives must stand or fall by its own strength. It is our strong hives that make our tremendous honey production and we never consider weakening strong colonies by taking frames of brood from them. And we are oftentimes startled by the way a seemingly weak hive suddenly booms out with a great number of young bees. The reason the hive appeared to be weak with few bees coming and going was that they were covering a large amount of brood, and when that brood emerges the hive almost miraculously passes from weak to strong. But if a weak hive does *not* respond to the increasingly warmer spring weather and build up in numbers, we let it go by the wayside to sometimes die out altogether. We want strong healthy blood lines in our bees. Natural forces will take care of the weak and sickly if we allow them to do so. We are, of course, running for maximum honey production.

Reusing an Old Hive

If one is planning to farm out his bees for pollination he might well want to even out the strength of each hive so as to have more of his hives sufficiently strong in bees to be acceptable to the farmer. But as far as the backlot beekeeper is concerned he will do well to encourage his strong hives to become stronger and, if necessary, replace a lost hive by cleaning and scorching out the old brood chamber over a bonfire or incinerator and hiving a new swarm. If we even slightly suspect that the original bees died out due to poison or disease we burn up the entire hive, top, bottom, brood chamber, contents and all. It is much cheaper in the long run to buy a brand new hive, buy or catch a new swarm and get honey, than to try to doctor an ailing colony for up to two years before the colony finally dies out anyway.

One afternoon a young man stopped to see me. It was obvious that he felt quite pleased with himself.

"How are your bees doing?" I asked him.

"Fine. A few days ago I looked completely through my strongest hive," he told me proudly.

"Why did you want to do that if you could see that it was really prospering?" I asked.

"I don't know," he replied. "I just thought it was the thing to do because everybody else does it."

I smiled. "Everybody except my father and me, maybe, but you'd never find us doing a useless thing like that. Needless rummaging through a beehive is sheer waste of the bees' time and energy, as well as a waste of their store of honey to reheat the hive. Did you learn anything?"

"Not really," he replied, "and before I was done I began to wonder myself why I was digging around in there so early in the season. I notice that one of your hives seems to be much weaker than the others."

"Yes. That hive is what we call a "slow starter" in that it always has a later buildup than the others, but even so by seasons end it makes as much honey as the others—sometimes a bit more. Some hives come out early and strong at the beginning of the season and stop work earlier too. But in any event it's safety on the part of the beekeeper to encourage his strong hives to become even stronger—for it's the strong hives that make an excess of honey and run our hive average up to around 150 to 200 pounds or more of wild flower honey per season."

Love Those Honeybees

Honeybee Quiz No. 2
TRUE OR FALSE. If you answer 10 of these questions correctly you know more than the average person. A score of 15 is excellent. *Answers follow the quiz.*

1. Honeybees will sometimes swarm and establish their new home in a fireplace chimney.   True____ False____.
2. In warm weather the queen and nurse bees suffer because the field bees are unable to carry water to the hive.
      True____ False____.
3. Bees can fly in a light rain if the weather is warm enough.
      True____ False____.
4. Honeybees gather nectar and pollen during the winter months in warmer climates.   True____ False____.
5. Due to the strong scent cedar lumber is unfit for building beehives.   True____ False____.
6. Redwood is an excellent construction material for beehives.
      True____ False____.
7. Drones are larger and heavier than worker bees.
      True____ False____.
8. Cotton is a nectar producing crop.   True____ False____.
9. Propolis is sometimes called bee glue. True____ False____.
10. There are only two species of hive bees.
      True____ False____.
11. All varieties of roses produce nectar.   True____ False____.
12. Location of the hive makes a great difference in honey production.   True____ False____.
13. Bees are color blind.   True____ False____.
14. Honey is lighter than water.   True____ False____.
15. Certain plants produce a nectar that the bees make into jade green honey.   True____ False____.
16. Bees often live in hollow trees.   True____ False____.

17. Beehives must be painted white to produce honey.

    True___ False___.

18. Worker bees live only six to eight weeks during the height of the honey flow. True___ False___.

19. A good place to locate an apiary would be beneath electrical high voltage lines. True___ False___.

20. Bees will visit plants with very small blossoms if they furnish an abundant source of nectar. True___ False___.

Answers:

1. True. And it is very difficult to smoke or burn them out.
2. False. When need arises they carry water in their honey sacs.
3. True. Our bees work steadily in warm light rains.
4. True. Honeybees in our warm area gather some nectar and pollen the year around.
5. False. Cedar lumber is excellent for building beehives due to its soft, stable, straight grain, and rot resistance.
6. True. Redwood is a marvelous construction lumber especially for small items like beehives, being durable and beautiful.
7. True. About one-half heavier and larger with square rear ends.
8. True. But the nectaries are under the leaf below the blossom rather than in the blossom itself. Strange.
9. True. Propolis is a very gummy glue, as all beekeepers know.
10. False. Honeybees fall into five main types, Italian, Carniolan, Caucasian, African, and hybrids.
11. False. So far as I have been able to observe only the wild five petal roses produce any amount of nectar and pollen.
12. True. If possible the hive should face south or east and receive some direct sunlight at least in the early part of the day.
13. False. Our bees like many different colors.

14. False. Water weighs almost 8 lbs. to the gallon, honey 12 lbs.
15. True. Sometimes we have had a little jade green honey but I have never been able to identify the source plants.
16. True. Hollow trees are natural habitats for honeybees.
17. False. Hollow trees are never white so we do not paint our hives white except to diminish the sun's heat.
18. True. But bees live longer than six to eight weeks in our cool moist climate. Sometimes they live an even shorter time where it is hot and dry or they have poor forage.
19. False. In our experience beehives placed beneath high voltage lines produce little or no honey, possibly due to the magnetic field.
20. True. Bees love our dry season tiny lippia blossoms as well as many other species.

7

Hive Preparation, Preservation, & Performance

The Problems of Painted Hives

"I SEE YOU'VE NO *white* hives," a visitor often remarks to me as he or she walks down the driveway and looks over the windbreak toward my beehives. "Why is that? Everyone else paints their beehives white."

"I know they do," I answer. "But my father and I don't for the very good reason that when we're trying for a world's production record in this cool area, we get more honey by painting our hives darker colors—if we paint them at all."

Then I go on to explain that we build most of our beehive boxes from redwood, a soft durable wood that is easy to fabricate. We rarely paint a beehive because we want the wood of the box to "breathe" as the bees work in it at the height of the honey flow. We want the excess moisture from the ripening nectar to pass out through the pores of the wooden sides of the hives rather than making it necessary for all of the water to be expelled by the bees through the hive entrance. Redwood is ideal for this purpose as are some other woods such as soft-grained pine. All of them should have some quarter-sawed or vertical grain in the boards so as to let the moisture pass through more easily. 🐝 We try not to use flat-grained boards for building either supers or hive bodies as such grain retards the passage of water through the pores of the wood.

Out in the sun redwood soon becomes a medium brown color which is pleasing to honeybees and it also seems to be able to absorb more heat than painted hives, especially those painted white. We do paint a few of our supers blue-gray, blue, blue-green, or dark red to give our bees some color variation to make it easier for them to recognize their hive from others close by on either side. But in a year or two the paint usually begins to peel off due to the interior moisture, and we rarely repaint. The point is that whatever we can do to help our bees, makes it just that much easier for them to make more honey for us. They are most willing to produce honey if we give them encouragement in a way that they can use.

In some areas of the world it is absolutely necessary to paint beehives even though doing so may cut down on total production. 🐝 In hot regions white or aluminum paint reflects the sun's rays and helps the bees keep the hive cool to the point where they can do their best work. If a hive becomes oppressively hot many field bees must be reassigned to duty as water carriers and this cuts down on the number available to gather nectar or pollen. Bees are as aware as we are that the fanning of hot air past cells containing water will cool a beehive, just as coolers use water filters to lower the temperature in our home. And of course in an overheated hive the water carriers have the added burden of carrying much more water for drinking purposes for use of the queen and her escort as well as the needs of the nurse and hive bees who are required to stay in the hive. If the overheating becomes too great we may see a large mass of bees gather outside the hive and cling to the area above the entrance or hang as a mass below the landing. 🐝 In this event our bees are not working at their maximum efficiency and they will be inclined to swarm if we do not help them by putting up shade, supplying a nearby source of drinking water, or painting our hives white.

Recently a young woman looked over our bee windbreak and remarked, "You'd have to paint all of your hives if you lived where we do!"

"And where is that?" I asked with interest.

"In the desert near Palm Springs, California," she replied. "The desert air out there is so excessively dry that all wood just seems to disintegrate into dust in two or three years—and so do our beehives!"

"I can belive that!" I exclaimed. "I've been in that dry heat too but didn't know that a board went to pieces in so short a time, but I too have seen boards all crumbly. How *do* you manage to keep bees out there?"

"We paint and replace as needed," she said, "and it isn't easy! We don't set any world's records but we do get a surprising amount of honey—considering the short time that flowers are in bloom."

"I'm glad to hear that," I said, "for I've never had an opportunity to keep bees in a desert area."

Circumstances do alter cases, as the old saying has it, and I am always delighted to find how an enterprising beekeeper can find a way to help his bees produce honey even in the less favorable parts of the world. Randle Brashear of Temple, Texas stated the matter succinctly when he wrote me a fan letter saying, "I can't set a new world's record in the area where I am but I do want to set a new production record for Bell County Texas!" My best wishes to Randle Brashear and all others like him.

Natural Interiors

Bees and oil do not mix. We never coat the inside of a new hive with linseed oil, or anything else, to preserve it. Often beginning beekeepers come to see me and tell of their frustrating experiences trying to hive a swarm of bees into such a linseed oil coated new hive. Sometimes the bees will enter the hive but they will almost always leave again within a day or two because they do not like the linseed oil treated boards. I do not know how to quickly remove the oil coating from a new hive for it soaks deeply into the soft grain of the wood. My only suggestion is to build a new hive body and discard the oil soaked one. If someone knows of a better way I would surely like to hear of it. 🐝 Actually, we do not need to try to preserve the inside of the hive as the bees do that themselves when really needed by coating the interior with propolis. And propolis is an excellent putty to use for filling small knotholes, cracks, and other imperfections in the lumber used for building beehives and related equipment. Even very dry and hard propolis will become somewhat soft and pliable if held in one's warm dry hands for a few minutes, after which it can be applied to the hole or crack as needed. And if at a future date the crack should reopen the bees can rework or add to the propolis to repair the damage.

Painting a beehive before bees have been hived into it can present a problem. This difficulty is of rather recent origin. In the old days we always used lead and linseed oil paint as that was the only kind of paint in general use for exterior farm needs. Our bees did not mind being hived into such a prepainted hive. But now with more sophisticated paints made from various kinds of new materials we find that our bees do not like the smell of these new paints and will sometimes resist our efforts to hive them. Recently we had a typical case in point. A man brought us a brand new beehive that he had just constructed. It was a nice piece of workmanship complete in every detail. It had a double coat of brilliant white paint all over the outside including the landing board. I immediately removed the hive cover and found that he had also painted the inside of the cover as well as all the interior walls of the hive and also the hive bottom all the way out through the entrance. I looked at his hive in dismay. Bees would not like his beautiful new thoroughly painted hive and I knew it.

"I brought you my new hive," he declared with a touch of

pride. "And I want you to hive a swarm of bees into it for me. What will you charge?"

"Eighteen dollars is our usual price," I told him, "and I'll try to get you a swarm but I won't guarantee it. I can see right now that no bees are going to like your beautiful box."

"Why not?" he exclaimed, "I've done everything right, haven't I? And a mighty good job of it too!"

"Your workmanship is excellent," I agreed. "But bees don't like modern day paint on their hive and certainly not all through the interior where they must live."

"But I wanted to preserve the wood of my hive," he stated a little crestfallen. "How do you lengthen the useful life of your hives?"

"I don't. You see we must make a choice between preserving our hives or encouraging our bees to make honey. My father and I want honey, and we are willing to replace our equipment after six years if need be. Most beekeepers depreciate their equipment on a ten year basis."

"Well do your best to hive me a swarm of bees," he urged.

A few days later I received a call from a homeowner on the west side of Santa Cruz. We loaded up the painted hive and drove across town to the address given. There we found a goodly sized swarm clustered on a small shrub. It was an easy hiving job as we could set our beehive on the ground and shake the bees into it from a low branch drawn over the top of the hive. The bees went right in—and then they came right out again—right out of the entrance, took wing and flew away to cluster on a small orange tree in the backyard of a home two lots away. We watched them go and then followed. I received permission from that homeowner to hive the bees and again we shook them into the hive and they sank down between the frames as at the first time. This time they stayed in for a few minutes but then they once more marched out of the entrance and flew away to land in a cluster a hundred yards away on a higher limb of an apple tree. As we watched them cluster for the third time my father warned me, "Now look, Ormond, we've already failed twice to hive those bees into that man's painted hive and the bees are getting nervous and flighty. If we want those bees at all we'd better hive them into one of our own new unpainted hives."

"You're right," I agreed. So I hurried to our station wagon and brought the hive we always carry with us for just such an emergency and set it up under the apple tree. The bees went in with evident goodwill as much as to say, "Now you're really buzzing on our wavelength. Why didn't you do this in the first place?"

As the days passed we tried several more times to induce a swarm of bees to stay in the man's thoroughly painted hive but none of them would do so. In the end he had me scrape and sand off all of the paint on the interior of the hive and then by coating much of the interior with beeswax we did, not without difficulty, induce a swarm to take possession of his hive and remain in it. I kept those bees at home for a week to make certain that they would stay in their hive. At first they seemed uneasy but after a few days they seemed more content to live in their new home and began to draw the starter sheets and the queen began to lay eggs in the partially drawn cells. I telephoned the man to come for his bees and everything ended well but at an unnecessary waste of time, effort, and money.

My father and I recommend that swarms be hived into un-painted hives and then ten days or two weeks later after the bees have become thoroughly established in their new home we take a paint brush and late in the evening after the bees are all "in bed" we paint the hive. The bees will not mind too much nor will they think of swarming out. They have a home and they will stay in it and care for their little ones—even as you and I would do.

Winter Wrappings

Rather than paint our beehives to help preserve them my father and I have, of late years, been wrapping them with sheet plastic. And, in addition, our experiments offer definite proof that bees in our wrapped hives come through the winter strong and healthy thus affording us a greater opportunity to break our own world's record of honey production. 🐝 We plan to wrap our hives on the evening of a quiet sunny day about November 1 and keep them wrapped until May 1 the next spring. Every four weeks during the time the hives are wrapped in plastic and even more often during rainy weather we need to unwrap the hives to see if the insulating hive covers are wet from condensation. If they are wet we dry them out or replace them with other dry insulating covers that we keep on hand for this purpose. We always unwrap our hives on a warm sunny day so that the outside boards of the hive can dry as well as the actual hive cover under the insulating cover. 🐝 If the sheet plastic is left on for longer than a month both covers often become moldy and there is a definite tendency for the wooden hive cover to dry rot. Thus it now seems that when using plastic wrappings to preserve the outside boards of our hives as well as to conserve the bees' natural hive heat we may have to replace the hive covers every three years due to the contained warm dampness trapped under the plastic. But with our honey production per hive so greatly increased we can well afford to replace the covers.

We do not know, as yet, why we find some hives to be sopping wet under the plastic whereas others are bone dry. It is time consuming to unwrap and rewrap a considerable number of hives. If we could ascertain in advance which ones were dry we would not have to expend needless effort unwrapping the dry ones. Two people can do a much faster job of wrapping hives than one working alone because the wind has a tendency to blow lightweight plastic in every direction before one person working alone can get a string completely around the hive to securely hold the plastic sheeting in place. But even so I have acquired considerable skill in doing it alone. 🐝 Since our honey flow begins here in Santa Cruz on or before January 1 each year it usually becomes necessary to add a first super to our strongest hives on January 30 and others at two to three week

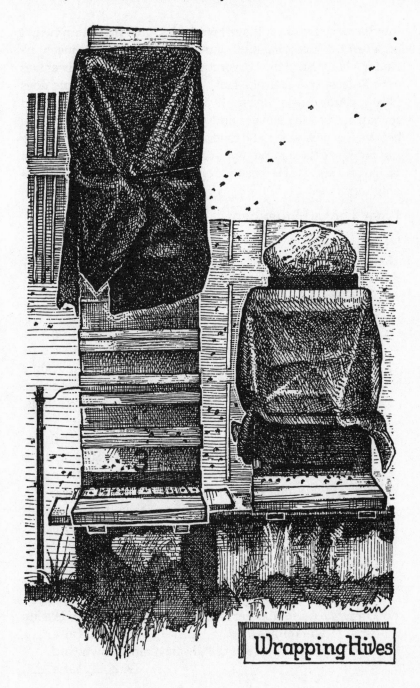

Wrapping Hives

intervals thereafter until our hives reach their maximum height of seven to eleven medium depth supers above the queen excluder about May 15. To vertically wrap such huge beehives with at least one and one-half turns of thin sheet plastic does pose a continuing problem. It does take labor to unwrap and rewrap each beehive every time we want to add a super but we believe it is well worth the time and effort. 🐝 As nearly as we can judge we have raised our average production thirty pounds per colony and that is incentive enough for us to continue wrapping our hives.

Love Those Honeybees

Honeybee Quiz No. 3
TRUE OR FALSE. If you answer half of the following twenty questions correctly you will know more than the average person. A score of 15 is excellent. *Answers follow the quiz.*

1. If a honeybee loses its wings it can grow a second set.
   True__ False____.
2. A honeybee can fly at a speed of forty miles an hour on a calm day. True____ False____.
3. A worker bee can carry a load as heavy as itself, such as, grasp the body of a dead bee and fly away with it.
   True____ False____.
4. In the coastal areas of California blue gum eucalyptus trees bloom from about January 1 to July 1. True____ False____.
5. Drones have long tapering tails. True____ False____.
6. It is difficult for a beginner to distinguish between a worker bee and a drone. True____ False____.
7. Queen excluders are used to confine the queen bee to the lower part of the hive. True____ False____.
8. From the time the egg is laid a fully mature drone emerges from the cell in 17 days. True____ False____.
9. Drones require a larger diameter cell in which to develop than do worker bees. True____ False____.
10. The opening to a queen cell always points downward.
    True____ False____.

11. During a mild windstorm honeybees are often shaken off from the blossom upon which they are working.

              True_____ False_____.

12. Rosemary has clusters of small blue blossoms.

              True_____ False_____.

13. In some areas bees gather a beautiful yellow pollen from morning glory, a species of bindweed. True_____ False_____.

14. Honeybees are all the same color. True_____ False_____.

15. All worker honeybees are the same size.

              True_____ False_____.

16. Honeybees are aware of variations in color.

              True_____ False_____.

17. When standing near his hives a beekeeper may be bumped head-on by a heavily laden bee coming in for a landing.

              True_____ False_____.

18. Drones help carry water to the hive. True_____ False_____.

19. Queen bees often leave the hive to get fresh air and enjoy the scenery. True_____ False_____.

20. Every day or so during the spring and summer the hive bees come out of their hive for a cleansing or "play flight."

              True_____ False_____.

Answers:

1. False. A honeybee has four wings, two on each side that hook together, but it must have all four wings in order to fly.

2. False. It flies at fifteen to twenty miles per hour, or usually a little faster than a fast sprinter can run.

3. True. It can even carry away a drone which is about one-third heavier than itself.

4. True. Of late years they have begun blooming as early as the middle of November due to our strangely warm weather pattern.

5. False. Drones have almost square chopped off rear ends.

6. False. Any beekeeper with good eyesight can readily see the difference.

7. True. Knowing for certain where the queen is makes it easier for a keeper to service his hives.

8. False. 24 days from egg to drone.
9. True. If you believe in keeping bees the way God meant them to be kept. Some beekeepers force drones to develop in worker cells, and they can, though not without stress.
10. True. No exceptions.
11. False. They have an uncanny ability to go right on with their work on a wildly blowing blossom.
12. True. Rosemary blossoms are usually excellent winter nectar producers.
13. True. And they are a joy to watch as they work in the two-inch long trumpet-shaped blossoms.
14. False. Bees vary in color from almost true black to almost true gold.
15. False. Honeybees grow for several days after emerging from their cells. Bees also vary in size from hive to hive.
16. True. In our experience blue-green wave cloths gentle bees more than any other color.
17. True. Often, yes, for the weary bees coming in for a landing seem to be less observant when near the home hive.
18. False. Drones do no work around the hive except to help keep the hive warm on frosty nights.
19. False. Queen bees seldom leave the hive though on occasion we do see one crawling around on the landing board close to the entrance.
20. True. It is sometimes difficult for a beginner to tell whether the bees have come out for a play flight or are going to swarm. During a play flight some bees always *return* to the entrance as well as leave the entrance. During swarming no bees return to the entrance until the swarm has flown away.

8

Requeening

Unfavorable Weather

Late in the afternoon of Monday March 21, 1977 Paul and Dorothy Ames stopped by to see us. They and their five children plus hired help operate Ames Apiaries, a large queen breeding layout at Arroyo Grande, California. The Ames specialize in a hybrid strain of black honeybees, as well as Caucasians and Carniolans. In addition to queen bees they also sell bees to start new colonies. Instead of selling two or three pounds as package bees they prefer to sell and ship what are called "nucs." These are two, three, or four drawn brood chamber frames of honey, pollen, eggs, young larvae, sealed brood, adhering worker bees, and a queen, all ready for insertion into a beekeeper's empty prepared hive. Such colonies build up quickly.

More than a year earlier the Ames had spoken to us of their marvelous new strain of honeybees and had strongly suggested that we try them out in our small experimental apiary here in Santa Cruz. However at that time it would have interrupted our experiments with our own bees so we declined. Now as the Ames drove up I wondered if I should ask them for one of their beautiful hybrid queens, but then promptly decided not to for the recent winter months had brought so little rain. California was in the midst of a drought. It seemed an inopportune time to try for a great production of honey either from our own bees or a queen from Ames Apiaries.

But almost immediately upon entering our house Dorothy Ames said quietly, "I'll give you one of our best queen bees— if you want her."

"Don't say that twice," I said as I looked her in the eye, "or I'll take you up on that offer."

Dorothy looked at me with a glow of true beekeeper's love in her eyes and said again, "I'll give you one of our best queens, a 'select tested' queen at that if you want her. Such a queen as the one I'm talking about would sell for twenty dollars, plus the cost of shipping and insurance."

For a moment I looked steadily at her and then at Paul.

"You heard what she said," Paul smiled as he spoke, "and she said it twice as you wanted."

"I'd love to have your queen!" I exclaimed.

"Then we'll send her right away," they said. "We've been keeping her in a holding cage for some months now."

"She's just what I'd want," I said as a tremor of excitement passed through me, "for I like an older queen for best maximum production."

And so the deal was made. Paul and Dorothy would send me a beautiful hybrid queen and as soon as possible after receiving her my father and I would introduce her into one of our already large beehives. We talked of many things that evening. Then the Ames left for home promising to package the queen and her escort and start them on their way to us by mail the next day so that we could introduce them into one of our colonies at the earliest possible moment for all of our hives were strong in numbers as our nectar flow had been on for more than six weeks. We already had storage supers on all of our hives. This would make

Aebi

Unfinished
Combs Due
To Drought

introducing a new queen more difficult but not yet impossible. I was glad the Ames had come to see us so early in the season for a month later when all of our hives would have from six to ten supers above the queen excluder I should have hesitated to attempt to requeen. 🐝 Requeening a strong healthy hive during the honey flow, from whatever cause or reason, always results in a lesser total production of honey from that hive for that particular season.

The beautiful black queen arrived by special delivery mail just before lunch on Friday, March 25. Ames Apiaries sent along with the queen a detailed set of instructions which my father and I read and studied until we knew exactly what we were supposed to do. Since methods of introduction vary considerably I will here give Ames' suggested method of introduction:

1. Remove the old queen from her colony or kill her, placing her in the bottom of the hive so the bees know their queen is dead and gather around her in a ball.
2. Keep both ends of the cage with the new queen and nurse bees corked or closed.
3. If no nurse bees have survived the shipping, go to your newly hatched baby bees or young nurse bees and add some to the cage (8 to 12 bees is sufficient). Make sure these young bees are full of honey so they cannot sting her. Smoking will usually cause them to store up. These nurse bees will readily accept the new queen and give her the hive odor by being in such close contact in the cage.
4. Put the double-corked cage in the center of the brood area at the top of the combs near the center of the frames, screen side down. Leave it for four days.
5. Four days later, return to the colony and cut out any queen cells to prevent supersedure. Uncap the candy end of the queen cage so that the bees can eat through and release her into the colony, return the cage to the colony, and leave her alone for five to seven days before examining the hive to see if she is laying and accepted.

These methods have proved successful for our customers and

ourselves in the past. Hope they will be helpful to you also. The instruction sheet also gives the following note of caution: Hybrid queens should never be clipped as this causes immediate supersedure in most cases! The bees think their queen is injured. (By clipping they mean snipping off the ends of the queen's wings so that she cannot fly and therefore must remain in the hive where she is placed.)

On the morning of March 25 a strong gusty wind from the north blew clear and cold—not far above freezing in the early morning hours. We fervently hoped that the wind would abate now that we had received delivery of our extraordinary queen. By eleven-thirty in the morning the temperature had risen to fifty-six degrees and my father and I were preparing to risk the cold and wind and introduce our new queen. But just then we had visitors, a young couple stopped by to talk bees and beekeeping. We took that as a good omen to refrain from trying to requeen during the cold wind. They stayed until one-thirty in the afternoon. By that time the wind had almost died down and the temperature had risen to sixty-eight degrees, even warmer in the sun out of the wind. The wind seldom dies down for long so we quickly set out everything we would need and got to work. We opened the hive, removed the two top supers above the queen excluder and set them aside. Now came the hard part for we would have to look through the entire brood chamber which was comprised of a full depth brood chamber and a medium depth super. We had been in the process of replacing the old full depth brood chamber with a new one due to the passage of time. We had already begun our five year exchange policy, so a new medium depth super with fully drawn combs from the year before was already in place immediately above the bottom board and the old full depth hive body was above the super, just below the queen excluder.

We had already removed the two honey storage supers so now we also removed the queen excluder. Quickly, due to the wind again arising, my father and I looked for the queen. We searched every frame, then brushed off all of the bees back into

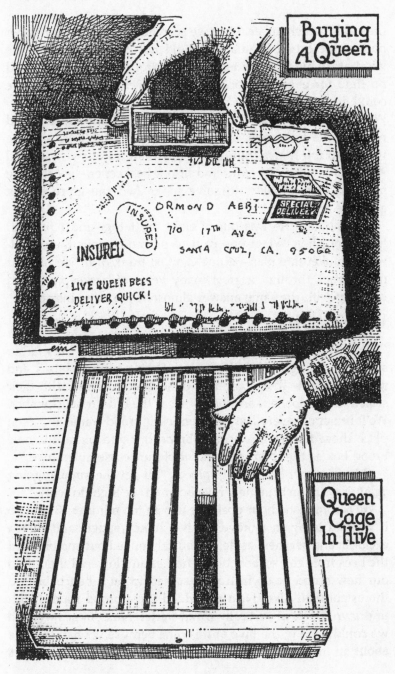

Buying
A Queen

INSURED
INSURED
INSUREL

ORMOND AEBI
710 17TH AVE.
SANTA CRUZ, CA. 95060

SPECIAL DELIVERY

LIVE QUEEN BEES
DELIVER QUICK!

Queen
Cage
In Hive

the old brood chamber. Then we searched the lower new super part of the brood chamber. We did not find the queen. The wind arose quickly, the temperature dropped rapidly. We had no time for even a second quick look for the queen. All we could do was reassemble the entire hive as quickly as possible. We were a bit dismayed—to say the least. Minutes after I had replaced the hive cover a young man came to talk bees so we went into the house. But after a few minutes our visitor, Paul White, a builder of homes and also a beekeeper, saw that we had our own problem to solve so he had the grace to say goodbye and leave.

The moment he was out of sight my father and I put our heads together. When one plan of procedure fails there is always an alternate plan—if one can just think of it. We ran our minds back through a great many years of beekeeping—and came up with the same answer almost at the same instant.

Dividing a Hive

"Forget all about finding and killing the hive queen," my father spoke *my* thought. "We'll get one of our new hive boxes and make up a new hive composed of six brood frames with their sealed brood, pollen, and honey from the old brood chamber. We'll brush off every last bee from each brood frame and then place those frames in our new brood box and place this new brood box as the topmost element of our reassembled hive."

"My thought, exactly," I agreed. "I'll get the smoker going again if you'll get the new hive box and we'll get to work."

We accomplished our goal in less than ten minutes. But to do it we had to again remove the two honey supers and queen excluder and set them aside. We quickly pulled out and brushed the bees from the six best brood frames and lowered them into our new brood box which we had set up on a bench behind the established hive. Next we added three brood frames with undrawn sheets of vertically wired starter. Nine frames were all we could have in the hive at this time because we had to leave about an inch of space between the center brood filled frames so that we could insert the queen cage. To accomplish this we wedged apart three of the brood frames from the other three brood frames and inserted the little wooden queen cage with our new queen in between them as per the instructions. Then we pressed the frames tightly together and against the queen cage by placing and wedging little blocks of wood between the outermost undrawn frame and the side of the hive body. This held the inserted queen cage firmly in place no matter how we moved or jarred the new hive box.

We rebuilt the hive as follows: Above the bottom board was the medium depth super part of the original brood chamber, next the full depth old brood chamber with six new frames of wired starter to replace the six frames we had removed, and then the queen excluder. Above that we replaced one medium depth honey super to supply the hive with food in case of inclement weather, and above that we placed another queen excluder. This second queen excluder was to give us added insurance that the established hive queen would not be able to find her way up into our newly assembled brood chamber which we now placed just above it with its six frames of beeless brood and our new queen in her cage.

Our theory of operation was simple. Heat rises, and bees will find and cover their brood no matter where it is in the hive—even if they have to pass through two queen excluders to reach their little ones. To aid in conserving heat we wrapped the entire hive in a tarpaulin and then in turn wrapped the hive with several thicknesses of heavy plastic.

The next day, Friday March 26, dawned sunny and warmer. After lunch we decided to separate the two brood chambers so we lifted and removed our newly made up brood chamber from off the top of the hive and placed it upon a newly prepared bottom board. It was now a complete one-full-depth brood chamber hive, and enough bees had gone up through the queen excluders to very adequately cover the six frames of brood. All was well.

We removed the upper queen excluder and placed a top on

the established hive. This hive now consisted of a bottom board, two medium depth supers for the replacement brood chamber, a queen excluder, the old brood chamber with four frames of brood and honey occupying the space on the warmest side of the hive, plus six undrawn frames, and the top cover. This completely reassembled hive we now took off from its stand and moved to another stand some sixty feet away around the corner of our house. Though the hive was very heavy because it was so big and still had most of the bees and much honey, we succeeded in moving it without difficulty by placing a circle of rope around the hive just under the lowest full width handholds of the lower medium depth part of the two-part brood chamber. Thus I was able to use my full strength to maximum advantage carrying on one side and my father on the other side. As we carried the beehive the floor board was only inches above the ground. It was easy to set the hive down so that we could rest, and then carry it on farther. When we had the established hive placed on the stand in its new location we hurried back and set the newly assembled hive on the stand exactly where the old hive had been. This gave our new hive with our expensive queen a great advantage in that all the field bees would return to the new hive in the old location. That is exactly what they did, and within two hours a great many had returned making our new hive very strong. But, as time passed, the established hive which we had moved away such a short distance became very weak as all but a few young hive and nurse bees left it to go back home to the old location.

By Monday noon, March 28, there were practically no bees to be seen coming or going from the moved established hive. This was due not only to the fact that so many bees had left it but also that the weather remained cold and the remaining bees had to stay indoors to cover as well as they could their four drawn frames only three of which actually had patches of brood. Our established hive was now far too large in cubic capacity for the bees to occupy and keep warm. We had known this would happen but we had been compelled to make the

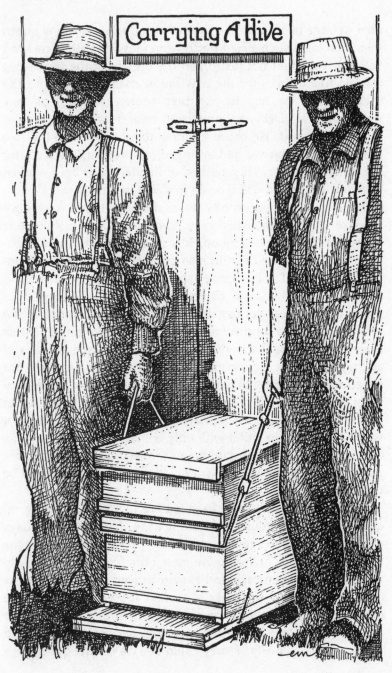

Carrying A Hive

hive so large because at the time of introducing our new queen there had been such a cold north wind blowing with increasing velocity. We could not just brush the bees from the frames we wanted free of bees for the new brood chamber, out onto the cold ground. We had to put them somewhere so that they would have at least some protection until they should leave of their own accord. We chose to brush them down into the two lowest supers that we had arranged, as previously stated, for the brood chamber. This method of handling our bees worked well and we would use it again in a like circumstance.

But at the first opportunity it was imperative that we reduce the size of our established hive. Our chance came about one o'clock on Monday afternoon, March 28. At that time we dismantled the hive and rebuilt it as a one full depth brood chamber having the aforementioned four frames of honey and brood plus six wired frames of starter yet to be drawn. Happily we found enough bees yet remaining to adequately cover all of the brood. Due to the continuing cold north wind we could not risk examining any of the frames to see if the old queen was still among them. But we fervently hoped that she was. If so, our experimental requeening and division might yet prove very satisfactory. Our established hive would be very weak for at least two months. When we began this requeening experiment we had expected to end it with only one hive, having killed the queen and replaced her with the new queen that had been sent us. But early in the season two hives are definitely more to be desired than one.

On that same Monday afternoon we opened the newly requeened hive. We found everything as we had left it four days earlier. The queen cage was still firmly pressed in between the two center frames of brood. We parted the frames and carefully removed the queen cage and turned it over. As many bees as could possibly get their heads around it were caring for the queen through the wire screen. I carefully brushed them aside and saw the queen inside the cage with her escort of bees and they were all as lively as they could be. We set the queen cage aside where the sun was warmest and proceeded to remove the frames of brood one by one and examined them carefully for signs of new queen cells either built or in the process of construction. We found five cells, three of which were about half completed and in which there were young larvae. Much as I hated to do so we had to destroy those queen cells with their larvae. This was necessary to assure acceptance of the new queen by the hive bees. We again looked for but did not find the old queen. This was a good signal for success. We removed the plug from the candy end of the queen cage and replaced the queen cage between its original two frames of brood so the hive bees could eat out the candy and release the queen within two to twelve hours. Then we closed the hive preparatory to waiting another six days before again checking the hive for a sight of our new queen, eggs, and young larvae.

During the days that followed I often thought of our two special hives. It seemed to me that our strong new hive to which all of the fielders had returned was well able to take care of itself. But I felt our previously established hive, still with only three frames of brood and one of honey, needed additional help. But what did it need? Then one night the Voice, of which I have spoken before in other of my writings, again prompted me. "Give those poor bees a drink of water," it said. That was all it said but that was enough to really set me to thinking. Water. We had had a recent rain, and dew every night, and none of the other hives had need for water because every morning their entrance boards were wet from the evaporation of nectar during the night. And I had not seen even one bee around our large pan of small stones and water that we used for a bee waterer. But then it occurred to me—our nucleus hive had no fielders—and I had noticed that their landing board was always bone dry. Truly, they needed water!

That morning as soon as the sun warmed the hive entrance I poured a few tablespoons of water on the landing board right in front of the tiny entrance. I had previously restricted the entrance to one-half inch high by five-eighths inch wide to conserve heat and prevent robbing. Within moments several bees came out and began sucking up the water as fast as they could. It took each one about a minute and a half to load up and crawl back into the hive. As each one left another came to take its place. They crawled excitedly around and some of them even buzzed their wings with joy. As they had need I gave them more water until at the end of half an hour they had all had a drink. Then for the first time some of them flew around the entrance to orient themselves to their new home and a few even flew away as fielders.

Strengthening the Divided Hive

Late that same afternoon Ronald Harmon, a high school student from Harbor High, came to see me.

"There's part of a swarm of bees on our athletic field," he said. "Could you come and hive them?"

"Yes," I answered. "Let's load up, and we'll go. You can ride with me."

In a few minutes we had loaded a hive, smoker, bee brush, and smoker fuel and were on our way. We found the poor bees on the ground near the dugout. The swarm had been run over by a land leveling device the evening before that had crushed most of the bees but there was still a large probably queenless double handful left as a little heap in the grass. I set our hive box close to one edge of them, started up my smoker and gently smoked them all around the edges of their pancake cluster. They aroused themselves and began to go into the hive. In fifteen minutes they were all inside. I placed the entrance closure screen and nailed it on. Ronald watched closely and helped where he could. When we had finished he said, "I've lost my fear of bees."

"Good!" I said. "But always remember that bees can sting, and keep a healthy respect for them at all times. When they do sting, other than when we're working with them in their hive, they're usually cross and out of sorts for a reason totally unknown to us. As a matter of fact we, as human beings, are sometimes cross ourselves—and without any obviously good reason—more often than not. Bees are much like people."

When Ronald and I arrived back at my home we combined that tiny remnant of a swarm with our needy hive. They immediately accepted the newcomers. All I had to do was brush the new bees off the combs in the catcher hive down onto the landing board in front of our weak hive. The newcomers crawled right into the hive and were accepted by the guards. This greatly strengthened our weak hive.

Three months later on July 1, 1977 I again checked both hives. Both were doing very well considering the exceptionally dry season. We had practically no rain during April, May, and June. Our beautiful black hybrid queen was flourishing in her hive with three medium depth supers above the queen excluder and all of them were becoming well filled with nectar and sealed honey. Our other hive, the other half of the original hive, was also doing well with two medium depth supers above the queen excluder. They too are now well filled with honey. Both hives should make excess honey for us the first year.

Love Those Honeybees

Honeybee Quiz No. 4
MULTIPLE-CHOICE. As there are no live bees in this book and you can't get stung for guessing wrong, try your brainpower on this quiz. A score of 9 will qualify you as a fair beginner in the life and lore of the honeybee. A score of 14 rates you as well informed, 16 is excellent. *Answers follow the quiz.*

1. Athletes eat pollen tablets because of their: High nectar content _____ Low sugar content _____ High protein values_____ Rich mineral content_____ .

2. Honeybees have: A single pair of wings_____ A double pair of wings _____ No wings at all _____ Wings of equal length_____ .

3. Sourwood honey is harvested from a tree that grows in: Montana_____ Nevada_____ Alaska_____ North Carolina_____ .

4. Which one of the following must have insect pollinators, such as honeybees, to produce a crop: Tomatoes_____ Carrots_____ Cucumbers_____ Lettuce_____ .

5. With a sufficient number of bees to properly do the job of pollination one of the following crops will produce a greatly increased yield of seed: Sweet corn_____ Alfalfa_____ Potatoes_____ Barley_____ .

6. One of these trees has blossoms and pollen that is poisonous to the brood of most honeybees: Apple_____ Madrone _____ Magnolia_____ Buckeye_____ .

7. Honeybees are greatly attracted to all but one of the following shrubs. Which is the exception: Hydrangea _____ Heather _____ Rosemary _____ Ceanothus (California wild lilac)_____ .

8. Upon the untimely death of the queen bee in a hive we still sometimes find newly laid eggs. These were laid by a: Young laying drone_____ A robber bee_____ Mature laying drone_____ A laying worker_____ .

9. In the Northern Hemisphere bees most often swarm in the: Spring and summer_____ Summer and fall_____ Fall and winter_____ Winter and spring_____ .

10. After emerging from their cells young worker honeybees usually remain in the hive for a period of: One day_____ Five days_____ Ten days_____ Two weeks_____ .

11. Honeybees have: One eye_____ Two eyes_____ Three eyes_____ Five eyes_____ .

12. One of the following trees is an excellent source of nectar: Locust_____ Sierra pine_____ Black walnut_____ Live oak_____ .

13. On a warm windy day honeybees fly: High up in the air to take advantage of the strong wind currents_____ Remain in their hives _____ Fly only short distances from their hives_____ Fly as close to the ground as practicable to minimize the detrimental force of the wind_____ .

14. In wooded areas propolis is obtained from: Ash trees_____ Pines____ Maples____ None of the aforementioned____ .

15. Drones mate with the queen while: In the hive_____ On the landing board_____ Both are clinging to a strong blade of grass_____ Flying 40 to 50 feet up in the air_____ .

16. Propolis is usually colored: Gray_____ Green_____ Reddish brown_____ Yellow_____ .

17. During the mating season drones may attempt to mate with: Only virgin queens from their own hive_____ Queens from nearby hives_____ Queens that are returning from a mating flight____ Any virgin queen in their flight area____ .

18. Honeybees dislike clothing made of: Woolen fiber_____ Nylon_____ Cotton_____ Linen_____ .

19. Sometime during the year our honeybees gather nectar or pollen from all but one of the following: Pussy willows _____ Black locust _____ Sweet clover _____ Peace roses_____ .

20. The queen may be distinguished from all other bees in the hive by her: Short dumpy body_____ Unusually long legs _____ Long and graceful body_____ Exceptionally large head and eyes_____ .

Answers:

1. Pollen has high protein value.
2. A double pair of wings.
3. North Carolina.
4. Cucumbers.
5. Alfalfa produces more seed with bees as pollinators.
6. Buckeye.
7. Hydrangea.
8. A laying worker.
9. Spring and summer are the swarming seasons.
10. Five days.
11. Five eyes.
12. Locust.
13. Bees fly as close to the ground as practicable on windy days.
14. Pines.

15. Flying 40 to 50 feet up in the air, mating takes place.
16. Propolis is reddish brown.
17. Drones mate with any virgin queen in their flight area.
18. Woolen fiber.
19. Peace roses.
20. Queen bees have a long and graceful body.

9

Our
World Record
Honey Producing
Hive No. 4

AFTER READING OUR FIRST BEE BOOK many of our visitors
have asked me what became of the beautiful queen of our world
record wild flower honey producing hive of 1974, and if we still
have that hive. Yes, we still have this hive which was once a
double-handful sized swarm that we called "tiny," and some
interesting things have happened to it. To fill in the story for
those who may not as yet have had an opportunity to read our
Art & Adventure of Beekeeping I will here again give some of the
early highlights.

On August 29, 1974 we took off the last excess honey from
our huge No. 4 hive. When we had it extracted and weighed,
and added this last amount of honey to all of that previously
bottled earlier in the season we found that the total weighed a
whopping 404 pounds. We knew we had set a new world's
record for wild flower honey! My father and I thanked God and
truly rejoiced that our bees had broken the old world's record
of 300 pounds set by A. I. Root of Medina, Ohio in 1895—and

our bees had surpassed it by a resounding 104 pounds! Furthermore, they had accomplished this amazing feat in their normal daily activity of life. Some have asked me if we used some special pressure on our bees to force them to produce so much honey, or if we used trickery of some kind. Mr. Root set his record in a forthright and honorable way and so did we. The only help we gave our bees was a lifelong expertise in knowing how to help them in their honey gathering, such as when to add or remove a super, plus a great amount of love. And during the nights of July 7, 8, and 9 of 1974 God sent us a most unusual two inches of rain that caused a myriad of nectar producing plants, white clover, lippia, and a host of others to spring forth as in a night and bloom profusely. Our hives being very strong in numbers the bees joyfully gathered a great quantity of this nectar which by August 29 they had converted into honey.

I immediately began to gather into a compact report all the facts, data, and pictures that would be needed to establish our claim to a new world's record with Guinness Book of World Records of London, England. I sent our report to Mr. Norris McWhirter, Editor of Guinness and he immediately replied that we had indeed set a new world's record for wild flower honey and that we would be in the next printing of their book of which he said they print 2,000,000 copies annually in fourteen languages. Our information reached Mr. McWhirter too late to be included in the 1975 edition but it was entered in their 1976 *Guinness Book of World Records*, page 98. As the months passed my father and I eagerly read everything we could find on honey production anywhere in the world to see if someone would quickly break our record. When the 1977 edition of Guinness arrived in one of our local bookstores I almost held my breath as I looked in the index under Insects for the greatest production, ascertained that the entry would be on page 101—and found that Ormond Aebi of Santa Cruz, California was still listed as the world's record holder! My relief was tremendous— and it hardly seemed possible that the name listed could be mine. We have lived so quietly, observing our bees and caring for them in the best possible way that we could learn from others or devise ourselves, that I had never really given much thought to being listed in any book, least of all Guinness.

Records are made to be broken and someone, somewhere, will break ours. And I will be glad for I am convinced that the ultimate in honey production has not yet been reached for no one as yet knows all there is to know about honeybees. My father and I carry on continuous experiments and in the 1976 season our No. 7 hive produced 310 pounds of honey again breaking the old world's record but not our own. Many visitors ask us, "Are you trying to break your own record?"

"Of course!" I reply. "For if we don't someone else will!" And we wish the blessing of God upon everyone who has the courage to try.

Inducing Swarming

Upon completion of my preliminary report on our No. 4 hive to Guinness I again began to give serious thought to our winning hive. Our beautiful queen was now almost past her third season of egg laying. She was getting old. In our experience she might last one more season but not longer. 🐝 The most logical procedure and the one taking the least amount of effort on our part, and also providing the greatest possible reward would be to induce our exceptional queen to order the building of swarm cells, the more the better, and induce swarming. If we used the other alternative in common use and removed four frames of bees, brood, and eggs from the brood nest, we could undoubtedly gain an increase of one. But if the bees would swarm naturally we might get a primary swarm and two or three afterswarms. And this was exactly what I wanted. But could we induce them to swarm? In our Santa Cruz area none of our hives had ever swarmed after July 15. Therefore there was a risk involved in trying to force our big hive to swarm so late in the season but my father and I decided to try it anyway as this had been one of my pet ideas for a long time. As the hive now stood

GUINNESS BOOK OF WORLD RECORDS

BY NOR

MORE AM

Hive #7

Again, Harry & Ormond surpassed A.I. Root's record 300 lbs. set in 1895

310 lbs of Honey in 1976

it had a brood chamber composed of two medium depth supers below the queen excluder and seven medium depth supers more or less empty of sealed honey except the well filled super directly above the queen excluder. The other six supers still contained small amounts of sealed honey and raw nectar in many of the frames. We saw no need to extract these last few pounds of honey because we had already far surpassed the old record, and also, the bees might need these stores for their immediate food supply if we had a prolonged rainy season in the weeks ahead.

What we needed to do was compress our bees into an ever diminishing space in the hive, and because of severe overcrowding, force them to build queen cells and swarm. It was a good idea, in theory at least, but it failed in actual practice. However, I could not see that far into the future so on September 5, 1974 I opened the hive and removed the top super leaving only six above the queen excluder. I would have removed more but the hive was full with upwards of 100,000 bees. This was a definite numerical decrease from the 135,000 bees that my father had conservatively estimated were in the hive at the height of the honey flow when we had nine supers above the queen excluder, but it was still a vast lot of bees. For the next few days from time to time during both day and night I pressed an ear against the side of the hive to listen in on the bees, but heard nothing to indicate that they were building queen cells. I discussed the problem with my father.

"Compress them even more," he advised.

"But how can I? The hive is full of bees and if I take off the top sixth super some bees will have to sleep outside tonight."

"Then let them sleep there," he said. "The weather is warm and they won't suffer."

I did as he suggested and that night a half gallon of bees clustered clinging in the open above the entrance. Again I watched and listened but still found no indication of preparations being made to swarm. Three days later on September 11, I again spoke to my father.

Our World Record Honey Producing Hive Nº4

Listening In On The Bees

"Our No. 4 hive gives no hope of swarming."

"Then remove the top fifth super."

"But if I do that some bees may have to hang out in a cluster under the landing board! Can we afford to take the risk?"

"We'll have to if you want to finish your experiment of forcing those bees to swarm to gain increase."

"All right," I said, "but this time I think I'll wear a veil as it's going to be a bit of a sticky problem getting the cover off, the bees smoked and brushed off the combs, that fifth super removed, and the cover replaced."

And so it proved to be. I think a few bees might have stung my face, had they been able, but none attempted to sting my hands. That evening a large mass of bees settled around and above the entrance and another cluster hung from the landing board.

"That's all the misery I'm going to give our poor world winning bees," I told my father that night. "Now they'll either swarm—or they won't. But they surely will—I hope!"

"It's your experiment," he said, "and I'll give you a 50–50 chance to win, but not more."

Eagerly, as the days and nights passed, I watched and listened for signs of swarming—but found none. September 1974 gave place to October and then to November, but never a sign of a swarm—and no longer any drones in any of our hives either. Now I hoped that the bees would not swarm for there were no drones to mate a virgin queen and if not mated within ten days after emerging a virgin queen becomes overage for laying the maximum number of eggs in the years to come.

"I think you failed this one, Ormie," my father said quietly one warm afternoon near the end of November.

"I think so too," I slowly admitted. "Our No. 4 hive still seems to be prospering, but that can be very deceiving, as we both know from experience. Have you any suggestions?"

"Not right now," he answered. "All we can do is wait and see how the bees come through the winter. A very mild winter would really help us."

Dividing Hive No. 4

A few weeks later in early January of 1975 my father had an idea. "Our No. 4 hive, as it stands, consists of three medium depth supers, two below the queen excluder and one above. Why not remove the queen excluder and in its place add another medium depth super containing ten drawn frames?"

"Why would you want to do that?" I asked. "Two medium depth supers have always been ample for a brood chamber— and now you want to double the queen's laying area by making it four supers high? Why?"

"Well just look at those bees work. Look at all that nectar and pollen they're carrying in. This mild weather caused the eucalyptus trees to begin blooming about December 1 this winter and in this past month all of our queens have begun laying eggs for the start of the spring buildup. As our No. 4 hive is especially strong why not double the capacity of the brood chamber and later gain increase by dividing it in the middle? The two top supers for one hive, the two bottom ones for a second hive."

"That's a brilliant idea!" I told him. "You get the super and I'll get our smoker, hive tool, bee brush, and a veil for you."

While my father was in the barn I had a chance to think over his idea. If left to themselves bees usually like to occupy a more or less egg-shaped brood nest in the center of the available space. If after removing the queen excluder we raised the top super and placed a super beneath it the queen would undoubtedly quickly move up into the drawn combs to lay more eggs. Then at a later date we could split the hive by taking off the two topmost supers and affixing a bottom under them thereby making a complete hive, and we could put a top board over the two lower supers thereby making that part into a second complete hive. Soon my father returned from the barn. In a few minutes we had removed the excluder and placed the additional super.

"By the way," I asked him, "when we come to dividing this hive, how will we know where the queen is? It's too cool to go rummaging through all of the frames to find her."

"We won't know—and we won't care," my father answered. "By the time we make the division the queen will have laid eggs in both parts. The supers with the queen will continue to build up rapidly. The ones without a queen will have to raise themselves another queen from an egg."

"It sounds logical and I hope you're right!" I nodded and exclaimed in pleased agreement. "It'll be a very interesting experiment."

The days seemed to zoom by like flying bees. The weather continued relatively clear and unseasonably warm. By January 16 our No. 4 hive was very strong considering the relatively small size of the overall hive. We counted as many as 130 to 140 bees a minute alighting on the landing board in the warm part of the afternoon all loaded with nectar or pollen. It was a beautiful sight.

"That hive has built up strong enough to split," my father stated with satisfaction as he surveyed it with a critical eye.

"Good!" I agreed. "Let's get our tools, a top and a bottom and get to work."

I had the smoker lighted and going well and was just in the act of prying the center supers apart when the Voice whispered, "Drones." Silently I thanked the Lord of Glory.

"Father!" I exclaimed. "There are no drones!"

"Say! That's right! Replace the insulation board, roof, and weights. We'll have to wait until we see drones. Came close to making a bad blunder that time."

"I wonder how long we'll have to wait," I mused aloud.

"For some time, probably. I forgot for the moment it's still midwinter. It's so warm it seems more like spring. Well, all we can do is wait."

So again we waited and watched. It was a very dry winter with little rain and not even much fog. The eucalyptus trees continued to bloom and our hives increased in strength every day. On February 14 a friend stopped by to see me.

"I saw drones in one of my hives two days ago."

"Good!" I exclaimed. "That's a really good sign. I just now saw a few drones come out of one of our hives."

"When are you going to split your No. 4?"

"On the next warm calm day."

February 16 was sunny and warm like a beautiful day in late spring. After lunch my father and I stood looking at our No. 4 hive. He caught my eye and nodded. I nodded in reply. I too had seen drones around the entrance of No. 4. Without loss of time we made the readjustment of the supers into two hives and placed the new hive close up against the original hive on the stand. We named our new topmost two-super hive No. 4, and located it just to the left of our original No. 4 hive which we now renamed No. 4½. The bees seemed to hardly notice what we had done. They flew away to the fields as before and returned as usual but we soon saw that all of the fielders from both hives were returning to our No. 4½ hive that was still in its original position on the stand. To rectify this situation we moved both hives eight inches to the right. That helped as some returning fielders entered our new No. 4, but not enough of them. So we moved both hives another three inches to the right and the bees entered each hive in almost exactly the same numbers. From time to time during the next six days we had to move both hives a few inches to either right or left to even the number of fielders in each hive, but we succeeded nicely. Then we moved the hives two feet away from each other on the stand so that the fielders would no longer cross from one hive to the other.

For the next two weeks both hives appeared to be almost equally strong. Both carried much pollen and nectar. At this point it was impossible to tell which hive had the queen and which was rearing a new queen. About March 5 the weather turned more windy and colder. Not really cold for our climate but more nearly like our usual late winter weather. Now No. 4½ continued to gain in numbers and No. 4 had noticeably fewer fielders at work. It was obvious that they were the ones undertaking to rear a new queen. By the end of March the field force of No. 4 had dwindled down until there were only thirty bees a minute coming in for a landing. We had been expecting something like this because the young bees reared the previous fall were now old bees who had been working very hard during the mild winter weather and could not be expected to live longer. For more than six weeks there had been no young bees emerging to replace the loss of the old. This was normal for it takes sixteen days for a queen bee to develop from an egg and a few days more for her to be mated and begin to lay eggs. And worker eggs take twenty-one days to emerge as fully developed bees and a few days more usually pass before we see them as fielders outside the hive. We had hoped the original queen would lay more eggs in the upper half of the hive that we had set aside as our new No. 4 before we separated the original hive into two parts. However we were glad that they at least had some eggs from which to rear a young queen.

By this time we had constricted the entrance to an opening only three inches long by three-eighths inch high. To conserve heat we had also wrapped the hive in black plastic sheeting over the top and down around all four sides until just above the entrance. First I wrapped the hive in two thicknesses of .004 plastic and then increased the wrappings to four, tying the plastic in place with two strong strings around the hive, one six inches down from the top, and the other around the hive just above the entrance. This lower string had to be placed carefully and drawn tightly so that no bees could crawl up under the plastic where they would be unable to find their way out again. This is especially true when using clear plastic for then bees will worry themselves to death trying to find an exit.

On April 10 we opened No. 4 and I removed three filled frames (ten pounds of honey) from the top super thinking the queen might be hindered in her egg laying for lack of empty cells. We replaced these filled frames with drawn extracted combs. It did seem that the young queen had been rather severely honeybound but until spring had really come I had not wanted to risk taking off the bees' stores too soon. During this trying period my father and I were highly tempted to combine a new swarm with our No. 4 hive. But that would have canceled our experiment and I wanted to see the final result even if it meant losing the hive altogether. On May 15 I was glad that we had waited for by that date the No. 4 field bees began to greatly increase in numbers.

On May 21 I again opened the hive and removed another seven pounds of honey to give the queen additional laying space. By June 15 there were up to 100 bees a minute coming in for a landing, but our honey flow was almost at an end. Yet even so, No. 4 managed to stockpile enough honey in their two-story medium depth super brood chamber to have ample stores for the coming winter. But they produced no excess honey for us as the seventeen pounds we had removed had been stored the previous summer by the bees in the original hive.

Hive No. 4½ on the other hand had built up strongly all season long. We placed empty supers on it as needed and by season's end it had produced 155 pounds of honey, by no means a record but still a goodly surplus of liquid gold. And in addition they had swarmed three times. This really pleased us for we so much wanted natural increase from our world's winning queen—and this time she did it! We kept one swarm for ourselves and sold the other two. The one we kept we used to requeen our No. 2 hive as those bees were just naturally too large to work through a queen excluder and we had been working them without one. We were completely successful as usual in combining our new swarm with the established No. 2 bees and so obtained a strong hive of our best stock.

As the winter of 1975-1976 approached my father and I were highly pleased with the prospect for our two special hives. Both had young queens. Hive No. 4 had reared a new queen and No. 4½ had obtained a young queen through swarming. Each hive had ample stores of honey and pollen to carry them through the winter plus a relatively large number of young worker bees to sustain the hive during the winter months. The outlook for the 1976 season could not have been better.

🐝 Again as usual on November 1, 1975 we gave each hive we owned a treatment of the drug Terramycin, the dose being one teaspoonful of powder mixed with one ounce of white powdered sugar poured into a small entrance feeder. A little pan capable of holding just one ounce of powdered sugar may be cut with an old scissors from the bottom of a disposable pie pan. The dimensions of the rectangular aluminum sheet are six inches long by two and one-half inches wide. Then by using a small piece of short straight board one can easily bend up the edge one-quarter inch all around the sides of the sheet. This makes the finished pan five and one-half inches long, two inches wide, and one-quarter inch deep. Filled level full it will hold just one ounce. Such a small feeder will slip into almost any hive entrance and can be pulled out again the next day with a foot-long length of wire bent one-eighth inch at one end and used as a hook to pull out the pan. One such feeding during the year is all we find it necessary to give our bees to provide the tonic they need to keep them healthy. Terramycin, usually available at feed stores, also helps prevent bee diseases, especially foul brood.

"But I don't believe in taking drugs of any kind myself, and I don't believe in giving drugs to my bees either," a visitor sometimes tells me.

"For many years I agreed with you," I answer.

"Then why do you give your bees a drug, Terramycin, now?"

"Because bees need our help to keep in good health so as to gather honey and do their work of pollination. Look at it this way. Satan tries to kill us. You know that. Think of all the accidents and diseases that befall mankind. Trouble, disease, and death everywhere. Satan also wants to hinder or kill our faithful hard working little friends the honeybees. But God is Love—and God counter-rules Satan by giving men the wisdom and knowledge to make life-saving drugs and medicines for both man and beast—and honeybees. I believe in using all of the life-saving remedies God has given us—including Terra-mycin for our bees. Leo Lazar, a drug salesman for Charles Pfizer Drug Company once told me this and I had to agree with him."

"I see what you mean," my visitor will usually agree after he has given the matter a little thought.

Truly we must help our honeybees in every possible way or they will lose the battle. Life is becoming increasingly hazardous and difficult for them too. I like to give our bees their feeding of Terramycin sometime during the heat of the day, usually around two o'clock in the afternoon when the bees are flying well and many of them are away in the fields. Those that remain in the hive pay little or no attention to me as I slip the small loaded pan into the entrance. Bees like the drug in its carrier of powdered sugar and go right to work licking it up and either eating it or storing it away in empty cells in the brood chamber. Years ago when Terramycin first became available we did as was recommended at that time, we opened the hive and spread the drug and powdered sugar over the ends of the brood chamber frames where the bees were supposed to find it and carry it down to store in the honeycombs. But too often the dampness in the hive hardened the sugar before the bees could store it away and as long as two years later we still found some of the mixture where we had poured it. Obviously such hardened drug was not doing the bees any appreciable good. So we experimented until we found a better way.

Two days after giving each hive its dose of drug I pull out each little pan. It is usually licked clean, but if not, its contents will be hardened by moisture in the hive to the con-sistency of rock candy so I take a small spoon or little block of wood and crush the hardened sugar into powder again and push it back into the hive entrance for the bees to work on for another two days. If at the end of the second period some powder still remains in the pan I recrush it yet again and give it back to the bees and during the following two days they will finish cleaning it up. But I always note the hives that empty their pans during the first two day period. Those are the hives that are in excellent condition for the coming winter. Hives that take six days or longer to complete taking the Terramycin may very well die out during the winter due to the failing health or actual loss of the queen. Just why this should be so I do not yet know but the results of our observations and experiments are leading us to this conclusion.

January of 1976 was again unusually mild and all of our hives built up strongly, three of them especially so, one of these being our new No. 4 hive. Therefore on January 30 we placed a medium depth super on each of these hives. It was a good move on our part for even though it was midwinter the bees handled the extra cold space very nicely and began to store nectar in it. It is risky placing supers so early because of the possibility of cold weather in February which could stop the queen from laying eggs for several weeks or even losing brood through chilling. If we see dead white larvae on the landing board in the morning it is almost always due either to the action of wax worms chewing their way through a brood frame thus destroying the cells and killing the larvae, or it is due to chilling.

As a rule if only one or two dead white larvae are found on the landing board early in the morning it is due to a wax worm. A dozen or more would almost always be due to chilling. Neither of the above is cause for worry but is an indication of what is going on in the hive. The bees can take care of the worms, and the chilling will not affect the hive too adversely unless the cold weather lasts too long or becomes too severe. In this case the queen may stop laying eggs for some weeks and we will have fewer fielders to gather the honey crop when the weather does moderate.

As the 1976 season advanced all of our hives showed promise of doing well. Our new No. 4 hive that had just lived, as it were, in 1975, now ended the season with a grand total production of wild flower honey of 264 pounds as excess for us. Our No. 4½ that had produced 155 pounds in 1975 again produced 157 pounds in 1976. We removed the last of the excess honey from No. 4½ on September 13. Soon we noticed that though the hive was still very strong in numbers it began to show signs of having a failing queen. This puzzled us for she was only one year old and should have been in the prime of life. 🐝 Nevertheless we knew she was failing because for some weeks we had seen no young bees out for air around the hive entrance. We could have requeened at that time but I wanted to further test my ability to recognize queenlessness without actually going into the brood chamber. During the first week in October the signs of an approaching queenless colony became more pronounced. In a way I was glad for there was another experiment that I wished to conduct and this bid fair to be an opportune occasion to try it.

May I Borrow an Egg?

In my research duirng the years I had found that on several occasions experimenters and close observers of honeybees had found that a hive of Caucasian bees on becoming queenless would sometimes send a worker to a nearby hive for the purpose of stealing an egg, which it could carry home to its own hive. Then the hive bees built a queen cell and raised themselves a new queen. But the Caucasians were the only bees reported to have this ability. For a period of years my father and I had a few hives of Caucasians and we had also observed this remarkable trait.

On one occasion in early spring we knew a large hive of our Caucasians was moping and had become queenless because they failed to fly strongly during the brief warm hours of the afternoon. We were highly tempted to try to give them a frame of brood and eggs from another hive, but the weather was not favorable. By the time the weather pattern had definitely cleared so that we could requeen with a minimum of risk we began to see discarded cell caps on the landing board of our queenless colony. With great interest we watched for a few more days and were rewarded by seeing many young bees around the entrance to the hive. Our Caucasians had requeened themselves! But they had stolen an egg from one of our Italian hives—and their new queen was an Italian, of course. So in a matter of weeks our Caucasian hive was converted into an Italian colony and in the end we lost all of our Caucasian hives for the same reason. We were unable to find a way to induce our Caucasian bees to steal an egg from a Caucasian hive so as to continue to propagate the species.

Now with our failing queen in No. 4½ I had an opportunity to try an interesting experiment. As far as I could ascertain no one had ever succeeded in forcing or coaxing a queenless Italian colony to steal an egg from any hive, Italian, Caucasian or any other and so requeen. Now I had an excellent opportunity to try the experiment. Our Italian hives Nos. 4 and 4½ were very closely related, having actually been part of the same hive only a year and a half earlier, and they were still located on the same stand almost side by side. Therefore I reasoned that if Italians would ever consider requeening with a stolen egg this was their best opportunity for as the strong colony diminished in numbers and the bees became more desperate for a queen it seemed reasonable to me to expect that if they could requeen by stealing an egg, that they would do so in their present dilemma. But as we found out later, they would not do it and this particular experiment failed.

During the November Terramycin feeding the bees of No. 4½ did not want to take their medicine—in fact it took them a whole week to finally finish and have the pan licked clean. We knew that their queen, though young and theoretically in the prime of life, was failing rapidly. Though it was very late in the season I might have requeened the hive had I not wanted to wait and see what the final natural result would be if I did not requeen. Slowly the many thousands of bees in the hive died and were dragged out of the hive and carried away by those yet remaining until on January 11, 1977 when I opened the hive the last few dozen remaining aged bees were all dead too. At no time during these months had there been any sign of disease, but an ominous sign of coming queenlessness which we had first observed and suspected on October 2.

Upon examining the frames I found large stores of both nectar and pollen, ample for the winter needs of the colony had they survived. I found no damage from wax moths for the weather had been just cool enough to keep the moth eggs from hatching. I found twenty-five partially completed queen cells, five of them of recent construction. Obviously at the time the queen cells were built the queen had been unable to supply even one of the cells with an egg so that the workers could save their hive by rearing a new queen. But in this instance the poor bees might have failed even if they had been able to rear a virgin queen because in the late fall of 1976 we had few drones to mate with her. A beekeeper one mile away also had a few drones around one of his hives, but the weather was uncertain and quite cool.

Why would a young queen fail in health and die like that? This is one of the mysteries we are still trying to solve. Fortunately it does not happen too often, and if recognized early enough one can requeen and so save the hive. We are thinking now of giving our Terramycin about October 1 of each year thus giving us a better chance to requeen any hive that does not quickly clean its entrance feeder pan, or plastic coffee can lid as some use, the dose being one-quarter cup of powdered sugar mixed with one teaspoon of the drug.

If I had not wanted to ascertain the final natural result of our experiment I would have requeened the hive immediately while it was still strong in numbers. I would have taken a frame of brood, eggs, and larvae from a strong hive close by and, after brushing off all adhering bees, placed the frame somewhere near the center of the brood chamber of the needy hive.

If a hive becomes queenless and short of bees before I notice it, as could be the case in an out apiary, I like to remove from an adjoining strong hive two frames of brood, eggs, and larvae with their adhering bees, checking carefully to see that the

Aebi

A Fine Hive
In Early Spring

established hive's queen is not among them, and introduce bees and all into the needy hive. In my experience the queenless bees joyfully accept the newcomers and soon raise themselves a new queen.

Caution: Before considering requeening in the fall and winter months with a frame and eggs be sure to observe other hives nearby to see if there is a good probability that there will be drones available for mating after the queen emerges.

As of January 20, 1977 our No. 4, the surviving half of the original split, was tremendously strong and it was mandatory that we place the first super of the season within a few days. If we failed to do this, whether it was midwinter or not, we would risk having the field bees bringing home so much nectar and pollen that they would fill too many of the available cells and so make our queen honeybound in the days to come. We had to make additional space available so that she would have a sufficient number of empty cells in which to lay eggs for the spring buildup. We were faced with a somewhat risky situation due to the future uncertainty of the weather, but we took the chance and added one medium depth super on January 30. On that date in 1978 the eucalyptus trees in the distance all around us were in partial bloom and the sun shone quite warmly—even though the news report said that it was freezing in Florida.

Love Those Honeybees

Honeybee Quiz No. 5
TRUE OR FALSE. Now that you have read this far in your book relax a little in a good comfortable easychair. Have a pen handy to check the answers, and prepare for some mental exercise that will be fun as well as instructive. Answer all of the questions even if you have to guess at a few of them. An unanswered question counts as an error. And you might guess right. Half of the 20 questions answered correctly is average, 15 is excellent. *Answers follow the quiz.*

All set? Go!

1. Worker bees are delighted if the nectar flow suddenly stops about two o'clock in the afternoon of a warm day.

   True____ False____.

2. During recent years many people in all walks of life have taken up beekeeping as a hobby or part time job.

   True____ False____.

3. In many areas much potential honey goes to waste each year for lack of beekeepers and bees to gather it.

   True____ False____.

4. When bees swarm they prefer to cluster in several small clusters rather than in one large one.   True____ False____.

5. As with human beings, the bodies of some bees are larger than others.   True____ False____.

6. The abdomen of an old bee appears to be more shiny than that of a young bee.   True____ False____.

7. Queen bees must be mated within 10 days after emerging from the cell to maximize their egg laying capacity.

   True____ False____.

8. As many as 7 to 10 medium depth supers may be needed at one time by a strong hive to enable the bees to gather and process the maximum amount of nectar available in the area.   True____ False____.

9. On cold spring mornings drones help to keep the hive warm.   True____ False____.

10. During the swarm season the first young queen emerging always kills all of the other young queens while they are still in their cells.   True____ False____.

11. During the mating season drones often fight and kill each other.   True____ False____.

12. Some bees are professional robbers rarely if ever trying to collect nectar or pollen but always trying to rob out a weaker hive.   True____ False____.

13. Often there are only 8 to 12 guard bees on duty at a given time near the entrance to a hive.   True____ False____.

14. A good queen is capable of laying 3000 or more eggs in one day, placing each egg in a separate cell.

   True____ False____.

15. When living in a tree in a forest heavily laden homecoming bees have great difficulty in finding their way home.

True_____ False_____.

16. Wax moth larvae sometimes destroy much good wax honeycomb.   True_____ False_____.

17. A disabled bee is nursed back to health.

True_____ False_____.

18. Ablebodied worker bees will feed starving worker bees to restore them to health and strength.   True_____ False_____.

19. If a worker bee drops onto a concrete surface from a height of more than 6 inches it is disabled for life.

True_____ False_____.

20. Queen bees may be severely injured if they drop as little as one inch onto a hard surface.   True_____ False_____.

Answers:

1. False. Bees often become cross if the nectar flow suddenly stops as they know that sunshiny harvesting hours are precious.

2. True. Those who come to visit us include medical doctors, nurses, clergymen, ranchers, housewives—truly every walk of life.

3. True. This is especially true in areas where large beekeeping operations are not allowed. We need many more backyard beekeepers.

4. False. Bees love to cling in one large mass. When we do find parts of the same cluster in several locations it is usually because the queen has alighted, left some queen scent, and moved to another branch of the tree.

5. True. Most bees can work through the standard queen excluder but almost every year we find one hive with bees too large so we work that hive without an excluder.

6. True. As a bee becomes older the fuzz on its back wears off and the bee appears darker and shinier.

7. True. Experiments have shown that early mating of the queen is highly desirable.

8. True. As nectar is largely water and it takes time for the bees to convert it into honey much storage space must be available to the bees.

9. True. Drones do eat a lot of honey but on frosty mornings they convert it into useful heat.

10. False. Young queens sometimes leave other queens safely in their cells and all but the last fly away as after-swarms from the hive.

11. False. I have never yet seen drones fight, period. They seem to be almost totally inoffensive.

12. True. Usually robber bees are not too numerous in our apiary but they are always most persistent and are a continuing problem.

13. True. Sometimes the scarcity of guard bees is what gives robber bees the chance to steal.

14. True. Queens that lay 3000 eggs a day are the ones that give a beekeeper a sporting chance to set a new world's honey production record.

15. False. They have a marvelous homing instinct.

16. True. Wax moths destroy much comb if the temperature in the storage area runs above 50 degrees F. Lower temperatures tend to inhibit hatching of wax moth eggs.

17. False. Never, I am sad to say, abandoned always.

18. True. It is a touching sight to see one bee feeding another.

19. False. It is beyond my comprehension how a worker bee can drop from a height of three feet and get right up and fly away—but they do it.

20. True. The egg laying organs of a queen are very delicate and we must not drop, or jar or squeeze a queen bee.

10

Preparing & Selling Honey— That Delicious Liquid Gold

Comb Honey

Aᴌᴍᴏsᴛ ᴇᴠᴇʀʏᴏɴᴇ likes to buy a little "honey in the comb" when they come to see us, as well as buy some of our extracted honey in pint jars (one and one-half pounds), and quart jars (three pounds). Comb honey sells for a good price and we are glad to cut it if we have plenty of newly drawn comb available. Most of the drawn combs we must extract so that we may return them to the bees to refill for a second or even a third time during the season. But often a strong hive will draw more comb than they need. If so, we take a lovely medium depth frame of honey, lay it over our uncapping box, cut it into four pieces of more or less equal size, hold a seven-inch paper plate under each piece as we cut around the chunk of honeycomb with a very thin sharp knife, and then drop it onto the paper plate. Later we slip the honey on its plate into a plastic bag, weigh it in ounces, staple our properly filled in sales label card to an end of the twistem used to close the plastic bag, and it is ready for sale.

But there are times when we are short of drawn combs to keep our bees working at their maximum efficiency and we cannot afford to cut out and sell so much newly drawn honeycomb. In that event my father and I cut what we call "window combs." We take a newly drawn and beautifully filled super frame and lay it over our uncapping box as usual, but instead of beginning at one end of the frame and cutting out consecutive pieces until all of the comb is removed, we usually cut out only three chunks. We make our first side cut about an inch from one end of the frame. This leaves a goodly amount of drawn comb with its honey to remain attached to the end of the frame. Our next cross cut is about four inches toward the other end of the frame. Then we make our end cuts, one along the bottom bar and the other along the frame back bar, but here again we leave at least one inch of honeycomb remaining attached to the aforementioned wooden parts. When our first cut-out comb falls into our waiting paper plate it leaves a "window" in the comb. Again we leave a full inch of honeycomb extending from the back of the frame to the bottom rail and make our next cross cut for the removal of the second comb of honey, or window, as it will be when this center part of the comb in the frame is dropped onto its plate. Once more we leave another full inch of honeycomb extending from the back bar to the bottom rail and make our third cross cut in preparation for cutting out and dropping the third and last piece of honeycomb.

Now as we hold up the frame in front of us using both hands, one at each end of the frame, we see that we indeed have a frame with three windows cut in it, and each of these windows is completely encircled by at least one inch of undisturbed, usually capped, honeycomb. Care must be taken to avoid breaking the two, one-inch wide center jambs that make the border for the middle window. We now place the window comb frame over a large pan to let the cut cells drain, and then we replace this frame in a needy hive as soon as practicable. The bees will quickly finish cleaning up the dripping honey, if any yet re-

Aebi

Cutting
window comb

mains, and will begin almost immediately to build new honey-comb in the windows to replace that which we have cut out. They love to begin work on such a comb and, if it is early in the season in our area, they will replace the missing comb so expertly that later it is difficult for us to find where we had cut out our three combs of honey.

Early in the season we sometimes have folks asking for comb honey before any frames are completely drawn, filled with honey, and sealed. If we find a beautiful partially drawn comb of honey with only one-half or two-thirds of the cells drawn and sealed on both sides of the starter sheet we cut such a frame into one or two window combs and return the unsealed portion to the bees.

Another advantage of cutting window combs is that we do not need to completely clean up the frame after the comb honey has been cut out and we do not have to replace the somewhat expensive starter sheet of pure beeswax. As beekeepers our labor and expense is considerably reduced and at the same time we are sometimes able to give our bees what almost amounts to drawn combs at a time when they need them the most.

Of course the day will come when these frames must be extracted for we can rarely cut window combs from the same frame more than two or three times because the bees so strengthen the comb we leave adhering all around the windows that it is hard to cut through the comb with our knife and also it becomes too tough for our customer to chew. Nevertheless, if done correctly, this is a simple and relatively easy way for beginning beekeepers to take honey from their hives without the need for an extractor.

In order to successfully cut such windows we must clean the honey and bits of wax from our thin knife blade after each cut through the honeycomb in order to minimize the drag of the knife through the fragile cells. This is easily accomplished by attaching a four-inch wide by ten-inch long piece of quarter-inch plywood to the far side of our uncapping box. Then using a hacksaw blade we cut a very narrow slot down an inch into the plywood and after every cut in tne honeycomb we draw our knife blade through this slot to quickly scrape it clean. The honey and bits of wax slowly drain down on the other side of the plywood into the uncapping pan.

Caution: Never try to cut comb honey from any frame in the brood chamber because the bees need those stores for their own use in brood rearing and also because most of the time the wax honeycomb is entirely too dark and hard for human consumption.

As can be expected the bees build the new comb to suit themselves and they usually build beautiful straight comb. However, it may be built on the pattern and size of worker comb, or it may have the larger drone type cells. On January 14 I was checking our supers for wax moths and found a lovely yellow extracted comb with three patches of drone cells evenly spaced with two vertical lines of worker comb between them. I was puzzled to know why the bees had built their comb in this unusual way until I held the frame up to the light of a sixty watt electric bulb and saw that it was one of my last season's window combs. Precious little bees had really used their imagination in redrawing that frame.

Finding New Markets

Selling the seasons crop of honey is always an enjoyable experience whether the sale is to people who come to see our operation or whether I am out peddling. I like to peddle part of our honey just for the joy of it as folks like to see a honey peddler come around. In peddling honey it is most important that people be able to recognize me from a considerable distance away. So I always wear the same clothing every time I go out, a brown felt hat and a green jacket. This combination of colors is visible for a long distance and is distinctive enough so that folks know it is I—the "honey man."

Main Street

On one occasion I was walking up Pacific Avenue, the main

street here in the city of Santa Cruz. I had been looking over the possibilities of selling honey on the lower end of the avenue near the beach, but not finding the prospects too good I walked rather hurriedly back up the avenue. On the way I saw a young woman on the other side of the street who seemed to be looking at me rather intently but I lost sight of her in the crowd of people passing on the sidewalk. So I kept walking and four long blocks later I heard far behind me a plaintive voice calling, "Honey man—wait!" Instantly I turned around and started back. After going back a block I saw the young woman coming toward me. She was pink cheeked and panting.

"Honey man," she said, "I saw you away back down the avenue but I was across the street from you. I thought I could catch you but you walked too fast and I kept getting farther and farther behind. So finally I called out—and you heard me! Do you have any honey to sell?"

"Yes!" I replied. "I surely do. My car is parked a couple of blocks away. If you want to stay here and catch your breath I'll be right back."

"Thank you, honey man. I'll be right here."

As I turned to leave I looked around at the small curious crowd that had gathered around us.

"I have bees and sell honey," I told them. "Anyone else want a quart or two?"

"Yes," several people responded, "bring me a jar too." But one elderly man continued to look at me with the most freezing stare that I have ever experienced. If I read his thoughts aright he was wondering what demoniac power I had used to cause that pretty young woman to come running after me—and what was worse—the other people present seemed to approve of her action and mine. As I think of it now I cannot help but wonder if that man was completely deaf, for when I returned he was not with the people who had waited for me to bring them some honey. To a deaf person our actions would have looked singularly suspicious. There is never a dull moment when peddling honey.

An Unpleasant Response

Do you always succeed in peddling honey to everyone you meet or are some people really "hard sell" characters? Unfortunately I do not always succeed in converting everyone I meet into an enthusiastic user of honey, as everyone should be. The above question always brings to mind my encounter with the Ogre of Cricket Hill, as I came to call him.

I was peddling honey on the far outskirts of my territory and had stopped to talk to some men who were standing around the door of a garage located in a small unincorporated village in the Santa Cruz mountains. As I came up and stated to all concerned that I had bees and therefore had honey to sell, and showed them my neat plastic wrapped honeycombs and also honey in quart jars, they all looked at me and shook their heads in the negative.

"But this is truly delicious honey," I told them. "You ought to be good to yourselves and buy some. Try it! You'll like it!"

But they all still shook their heads in a negative reply, and glanced back toward the farther part of the garage.

"You'll have to talk to the boss," one of them finally said.

"Who's the boss?" I asked.

"The tall gray haired man in the dirty white shirt," one of them volunteered.

With my arms loaded with two carriers of honey I went over to see the man.

"If you have a minute, please look at my honey—."

"I don't have a minute, and I don't want any honey, and I don't want to talk to you, and I don't want to see you in here—and so—get out!"

As I opened my mouth to reply he began all over again and stepped toward me until he was so close that I had to step back to keep him from stepping on my toes. Every time he paused for breath and I opened my mouth to speak he began all over again with a tremendous torrent of words—and all of it meaning "out!"

I even tried to offer him a free sample jar of honey or a big bite of honeycomb but all the response I got was to have him threaten to take a bite out of me! Such a carnivorous looking character I had never met before—nor any like him since. He backed me out of the door to the great amusement of all those watching. Then I realized that they all knew this man and had known what would happen if I tried to talk to him. Some of them had met me before and they knew that I had a pretty good line or "pitch" for selling honey and they were eager to see how I would handle this man. Well I could not handle him. He could out talk me two to one. So with my honey in hand I backed out of the door and down the driveway as best I could and then kept on going when he finally stopped crowding me. About ten feet away from him I stopped and looked at the other men and laughed with them. We must have put on a truly comical exhibition. And with the laugh came a happy thought on which to leave so I said, "Mister, if you'd take it I'd still love to give you a pint of honey—it would do wonders to sweeten your disposition!" And I meant it. He flashed me a strange quizzical glance, then without a word turned and went back into his garage. The men laughed again but this time they were on my side and one of them even inquired about buying a little honey.

Wealth versus Wisdom

Rich men interest me. It has been my good fortune in life to work for periods of time with truly wealthy men—millionaires, really. Such men are demanding. They want what they want when they want it. But they are also generous in many ways and pay the man well who can accomplish what they want done. Some years ago a wealthy man who had made his fortune in a large city came and bought several parcels of land in our area. One of these was a wooded acreage on the far edge of my honey peddling territory and on it the man was building a large new home. I made a routine stop at his construction site one day and had the good fortune to meet him in person. He was out in the hot sun shoveling fresh concrete in a manner far beyond his strength but he said he was enjoying himself tremendously. He immediately bought a dozen pint jars of honey and some honeycombs, and invited me to stop again in one month. This I agreed to do. But before the month had passed he telephoned me that he wanted more honey and he wanted it by two o'clock that afternoon. I told him that I would have it at his place promptly at two.

"But you can't make such a trip pay," my mother remonstrated. "Even if he buys another dozen pints of honey, that will no more than pay for your gasoline for such a long trip. And it completely upsets our own work schedule."

"You're right," I told her, "but I'll also take along a couple of dozen quart jars of honey and make a number of stops along the way. I'll sell enough honey to make the trip pay—and we can rearrange our work schedule. It doesn't really matter whether I pack honey in the daytime or at night." And to this my mother agreed.

That summer "Mister Rich," as we came to call him, bought a surprising quantity of honey. He often gave great parties for his city friends and he always featured our honey, using it for prizes and in various other ways. We had many hives at that time and much honey. The free advertising he gave us that year was worth a small fortune to us as our honey and name became known far and wide. One day when I had finished making a delivery he asked me to stay a few minutes longer to talk.

"I want to buy one of your beehives," he said. "I want you to bring it to me tomorrow at eleven in the morning and I want you to set it up on the top of a ten foot pole that I'll have set up by that time. And I want you to come over every Thursday afternoon and drain out the honey so that I can have fresh honey every week."

"Where will the pole be placed?" I asked in amazement.

"Right there on that sloping creek bank," he answered.

"How big a pole are you planning to have?" I asked in mounting consternation as I realized the utter impossibility of complying with his request.

"Oh, a four-by-four timber or maybe a six-by-six inch timber sticking up ten feet. Why?"

"I can't place a beehive up on a pole like that," I said. "And even if I could, I couldn't work it to take off the honey."

"Why not? I shall expect you to install a spigot at the bottom of the hive. All you'll have to do is open it and drain off the quantity of honey I require."

"Beehives don't work that way," I told him. "I can't possibly do as you ask."

"I have money," he said, "and money can buy anything."

"Anything, maybe, except a beehive with a spigot on a pole! You are in a marvelous honey producing area with eucalyptus trees and ceanothus in great abundance and if I were to work your bees to their maximum potential, during the spring and summer I'd have to stack supers on your beehive to a height of more than six feet. And frames of honeycomb must be removed from the hive, uncapped and extracted before the honey can be drained into jars or other containers. Sir, do please drive over and see my bees. Then you'll be able to understand the impossibility of what you ask."

He turned and walked away, and I drove home. I thought he would never phone me again, but he did.

"Are you going to bring me that beehive?" he asked.

"No," I told him. "Come see my bees and you'll understand my predicament."

A few weeks later he did come to see me, driving the most luxurious automobile I had ever seen. He stepped from his vehicle and walked with me to our nearest huge beehives on their stands.

"Oh!" he said as he saw them. And that was all that he said. He turned around and drove away without saying another word.

"Mister Rich" continued to buy honey from me for the balance of the season. I liked that man and was sorry when I learned that he had suffered a heart attack and passed away. Rich men are rich because they are exceptionally well informed in their own line of work, and they also have the ability to hire better men than they are to work for them. But when they try to enter a new field of knowledge and labor they need basic instruction, assistance, and encouragement even as you and I.

Greed Spoiled a Good Thing

My friend Bill lived for many years in the San Joaquin Valley of California. He owned a large diversified ranch and among other things kept 200 hives of honeybees. He told me they did well for him though he never had time to work them to their maximum ability. His neighbor also kept between 200 and 300 hives and he and his neighbor traded work when extracting time came around. Then his longtime neighbor sold out and a younger man from a small distant town took over. In due time this new neighbor, with Bill's help, learned to harvest at least a modest amount of honey from the 300 hives of bees that were included as part of the purchase of the ranch.

One hot summer day as they finished extracting Bill's honey his neighbor said to him, "Bill, your 500 gallon tank truck isn't quite full of honey. Why don't you just fill it up the rest of the way with water? Honey and water mix well and by the time you get to Fresno the packer there will never know the difference."

"Not me!" said my friend Bill. "Don't kid yourself, they'll know the difference." Bill knew that bees carefully control the water content of honey to preserve its sweet and lasting goodness.

Some weeks later Bill went over to help his neighbor extract. When they were finished and had run the honey into his neighbor's tank truck, again the tank was not quite full.

"I'm going to add twenty gallons of water to that tank of honey," his neighbor remarked. "I'm not going to drive all that way without a full load. They'll never know."

"Don't be a fool," Bill warned him again. "They'll know, all right." But nothing Bill could say or do would deter his greenhorn neighbor from pouring water into the tank on top of his good honey. Late that afternoon Bill's neighbor returned from unloading his truck at the packing plant.

"They never knew the difference!" he boasted. "I just drove up to the unloading area, they asked me the name of the ranch I was from, and we unloaded. Bill, that's a slick way to make a few bucks, only you don't have the nerve to try it," and he laughed. Bill said nothing for he realized that his neighbor had succeeded in the swindle because of the good name of the ranch. But he knew the packer would be on the lookout for his neighbor the next time he brought in a load of honey.

Weeks later, during the hottest part of the summer, Bill's neighbor extracted again, and again the tank was not quite full.

"I'm going to put in about thirty gallons of water this time," he remarked. "They'll never know the difference this time either."

Bill said nothing.

"By the way," his neighbor called out as Bill was heading for home, "why don't you go to town with me and see how this honey business is done?"

"I think I will," Bill answered, "and since I want to go to town this afternoon I'll drive my car and follow you."

"Good enough," responded his neighbor. "I'll meet you there."

About an hour later Bill pulled up at the packing plant. His neighbor's rig was out in the parking lot.

"Unloaded already?" Bill asked.

"No. When I called in and gave the name of my ranch they said I'd have to park my truck out in the lot as they couldn't handle my honey right away. They'd let me know when my truck was unloaded and I could come and pick it up. It's late now so they won't unload tonight. May I ride home with you?"

"Sure," Bill replied. "Hop in."

The following three days were hot and the nights were hot too. Late on the afternoon of the third day Bill's neighbor came over to see him.

"Got a call from the packing plant," he said. "They say my

honey spoiled out on the lot and I must come and get it. Will you take me to town?"

"Be glad to."

Upon arrival at the plant Bill's neighbor went to talk to the man in charge of unloading.

"Why didn't you unload my honey right away? You didn't have to leave it out in the sun to spoil. What made it spoil anyway?"

"Too juicy, probably," the man answered. "The boss said to leave it out on the lot and that's where we left it."

"Well what'll you give for it now?"

"Not a cent! Nothing! But get it out of here!"

"What'll I do with it?"

"You might try feeding it to your hogs if you have that many," was his acid reply.

Bill's neighbor silently climbed into his truck and drove home. Then he dug a deep hole and poured the entire contents of his honey tank into it and covered it up for a total loss of approximately 450 gallons.

"Bill," I asked, "how did the honey packer people know your neighbor had watered his honey?"

"I don't know that either," he replied, "but stop to think a moment. Common sense would tell us that they must have some simple and easy way to test for water in honey. Otherwise every shady minded character from here to Texas would be bringing in watered honey to sell for nine cents a pound, the same as we got that year for honey in bulk in the truck." I nodded in agreement.

"You're right, Bill, of course."

"My neighbor could have sold that truckload of honey for about $465 but he added water worth about $22 and spoiled the whole lot. How dumb can one get!" I quite agreed with him.

Samples Make Sales

It is always a joy to meet visitors who really want to know more about bees and honey. As I walk down our driveway toward them I am rarely successful in guessing whether they already have bees of their own, wish to acquire bees, or just want to talk bees. Some of course, come to buy samples of our bee equipment or honey.

"That package of comb honey looks positively delicious—but how do you eat it?" My questioner was an unusually tall and beautiful young woman named Catherine Banghart who, followed by her equally tall and handsome husband David, had upon my invitation entered with me into our honey house and extractor room. What momentarily startled me was the fact that she had asked exactly the same question that so many other visitors have voiced and I wondered for a moment if she was trying to kid me. But an upward glance at her face assured me that she really wanted to know.

"Most folk," I replied, "just take a silver table knife (more often stainless steel these days) and cut off from the square of honeycomb a generous bite-sized chunk sufficient to fill their mouth. This they chew for ten or fifteen minutes until all of the honey is squeezed out of the comb and the residue has become a small, usually compact, wad of wax. This they discard and help themselves to another bite-sized chunk. This is the way I eat comb honey. Two helpings are all one should eat at one time for honey is a highly concentrated food.

"But some of our elderly citizens crush as much of the honeycomb as will be needed to spread a slice of bread and eat it honey, wax, and all—and they insist that this is the only true way to eat comb honey."

"We'll eat it your way!" my visitors exclaimed in unison as they proceeded to cut for themselves a helping from the honeycomb I offered them. "It's so good!"

"And so good for us healthwise," I added. "Also, local honeycomb with pollen brings great relief to many who suffer from an allergy such as hay fever."

My visitors then purchased several lovely white wax capped combs, beautiful examples of the honeybees' exquisite craftsmanship in comb building.

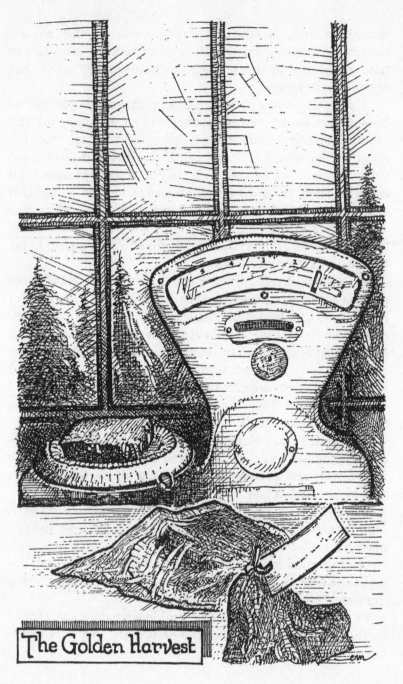

The Golden Harvest

Recently a chubby five year old friend of mine followed me into our honey extractor room where I had been working. When he saw all the beautiful small discarded pieces of honeycomb in the uncapping box his eyes widened as he looked up at me.

"You may have all you can put in your mouth," I told him, "but only one filling."

After a moment or two I realized that those were not quite the right instructions to give to a five year old, for he began stuffing the dripping honeycomb into his mouth until his mouth was full to overflowing. Then he gave me the most satisfied glance that I have ever seen on a child's face, turned and went back outside. For the next twenty minutes he walked slowly around the yard in complete silence, all the time endeavoring to close his mouth sufficiently enough to chew. Of course all of us who watched were highly amused but none dared to laugh aloud as that could have started our young friend to laughing too and he might have choked on his supersized bite of honeycomb—and to strangle on honey is a serious matter.

But in due time he finished his honey without mishap and came back to where I was working exhibiting the most satisfied expression imaginable on his thoroughly gooey and besmeared little face. We took him over to the hose, and honey being readily diluted and removed with water, we soon had him all cleaned up and the warm sun had his face and clothes dry in a few minutes. For a hungry little boy or a quick snack for any of us one cannot do better than enjoy a little honey.

The Beekeeper's Delight

Honey in the jar
　Honey in the comb
　　Honey in the kitchen
　　　Let's eat at home!

Honey sweet honey I spread on my bread
　Pour it on cereal or crackers instead.

Like it in applesauce, hot tea or ice cream
　Love it in coffee, it's really a dream.

Give me a spoonful, one more my dear
　No need to be skimpy, we've enough for a year.
One cannot beat honey for flavor and health
　For honey gives longevity, sweetness, and wealth!

　　Queen in the palace
　　　King on the throne
　　　　Never lived better
　　　　　Than with Honey at home.

Love Those Honeybees

Honeybee Quiz No. 6
TRUE OR FALSE. In answer to the accusation, "You ask such hard questions," may I say that all questions are hard if one does not know the answers—and all questions are easy if one does. So think positively and again come out a winner on this one. The usual rules hold. Ten answered correctly is average, 15 is excellent. *Answers follow the quiz.*

1. A swarm of bees will remain clustered on a tree limb for a definite length of time.　　　True__ False____.
2. Scout bees are sent out to look for a permanent home for the clustered bees.　　　True____ False____.
3. Worker bees that draw the honeycomb always work together in a cooperative effort.　　　True____ False____.
4. Drawn honeycomb in a hive darkens with age as the years pass.
　　　　　　　True____ False____.
5. Bees clean each cell before refilling it with honey.
　　　　　　　True____ False____.
6. Bees like to crawl head first into an empty cell and lie in a horizontal position while sleeping.　　　True____ False____.
7. Even with smoke and a bee brush it may take a minute or

longer to arouse a soundly sleeping bee so that it will crawl out of its cell. True____ False____.

8. The worker bees care for all of the personal needs of the queen. True____ False____.
9. All honeys have the same flavor. True____ False____.
10. All hives produce an approximately equal amount of honey. True____ False____.
11. There are effective remedies for bee stings. True____ False____.
12. Because bees pollinate the major legume crops, bees are of prime importance in the production of an adequate supply of meat. True____ False____.
13. All of the facts regarding honeybees are fully known. True____ False
14. The overall height of a beehive should never be more than four feet. True____ False____.
15. In the United States all beehive supers are standardized to one size and shape. True____ False____.
16. It is possible to remove a honeybee from a windowpane with one's bare hands without getting stung, or injuring the bee. True____ False____.
17. The honey flow can occur at any season of the year depending upon the geographical location. True____ False____.
18. The young larvae in a beehive look like small white worms. True____ False____.
19. Queen cells are always found near the top bar of a frame. True____ False____.
20. When we take honey from a hive we may find bees asleep in empty cells. True____ False____.

Answers:
1. False. Swarms are totally unpredictable, any length of time from fifteen minutes to two weeks, so hive bees as soon as possible.

2. True. At first only two or three bees may fuss around the entrance of an empty hive, then increasingly more as the days pass.

3. True. We often see "ropes" of bees working together to draw honeycomb particularly if there are no starter sheets.

4. True. Combs darken due to being often revarnished.

5. True. Bees thoroughly clean each cell and usually revarnish it before refilling with honey.

6. True. Bees can sleep in any position if need arises but they prefer to sleep in much the same position as we do.

7. True. Sweep the brush over the tails of sleeping bees and then allow them a minute to awaken and crawl from their cells.

8. True. The queen almost always has an escort that cares for all of her personal needs.

9. False. Honeys vary greatly in flavor depending upon the source of the nectars.

10. False. Honey production may vary greatly from hive to hive due to wind and hive location, strength of the hive, swarming, health of the queen, and the experience and know-how of the beekeeper tending the hives.

11. True. An excellent remedy is Sting Kill. I have just been told that onion juice applied to the sting is also effective for some people.

12. True. If it were not for the beautiful cooperation between honeybees and legume crops we would have little meat to eat.

13. False. Beekeepers the world around are experimenting to learn more of the true facts relating to honeybees.

14. False. The height of a beehive depends upon the abundance of nectar in a given area during the main honey flow.

15. False. Some beekeepers still use eight frame hives and supers, and others ten frame, and then there are all kinds of experimental sizes and shapes.

16. True. By exercising skill and care anyone can remove a honeybee from a windowpane barehanded.
17. True. Northern Hemisphere honey flow is during the spring and summer, Southern Hemisphere, our fall and winter.
18. True. Precious tiny white worms at first which quickly grow into much larger somewhat yellowish worms just before going into the pupa stage.
19. False. Swarm cells are found along the bottom rail of a frame or frames, supersedure cells halfway up in special cells pointing downward built into the face of the drawn comb. Bees build a supersedure cell when their queen is failing. We never destroy a supersedure cell as it may well be the hive's last natural chance to rear a new queen and survive. Replace that comb with great care, especially if it is sealed!
20. True. They may be sleeping lightly or soundly and we must awaken them with loving consideration.

11

The Language Of Bees

H<small>ONEYBEES</small> have a fascinating language that every bee-keeper should attempt to learn as quickly as possible. It is composed of very high pitched sounds, motions, and pantomime. As with human beings some bees talk more with their motions than with their voices. I have always found it most interesting to watch certain people who are in earnest conversation. How their hands and arms do gesticulate and their bodies gyrate! So it is with bees. We can determine from watching our bees, at or near their entrance, many things about what they are doing within the hive without ever removing the cover. Every time we open a hive we are interfering with the bees' carefully organized work routine and each disruption always results in less honey production. So the goal of every bee-keeper should be to learn their language sufficiently to enable all of us to know what we can do to encourage our bees to work to their maximum efficiency, which is just another way of saying that we want to help them to enjoy their life to the full.

Bees work very hard during the season of honey harvest but they enjoy every minute of it. And as beekeepers we can derive much added enjoyment from watching our bees as they come and go or fan happily on the landing board after we have learned their language to the point where we can understand what help they would desire from us.

On the morning of April 22, 1977 Father Markey stopped to see me. While I was yet in the house he paused to observe our hive No. 4 and saw what appeared to be a small dead bee on my plywood ground cover board below the extended landing board. While he watched, another bee flew down from the landing, walked in a circle around the motionless bee and seemed to be talking to it. In a few moments the apparently dead bee began to move and come to life. When I came out of the house we watched it together. We saw the hive bee position itself mouth to mouth with the weak bee. In a minute or two the weak bee was noticeably stronger and more active and soon began to crawl around with obvious delight.

"Why did they do that?" Father Markey asked with great interest.

"The bee on the ground board was hungry and unable to fly up to the hive entrance and the hive bee just fed it. Probably the weak field bee had been out late last evening and couldn't make it quite all of the way home so it had to stay outside last night. And now that the hive bee has fed it, in fifteen minutes or so the weak bee will be strong enough to fly up to the hive entrance and again join the work force." And that is exactly what happened. Bees often feed starving sisters, but they never feed nor care for sick, disabled, or aged bees who are unable to be of further use to the hive as a whole.

Every spring we see the "bee dance" which has been much discussed by other writers. This dance, by a returned field worker, shows the other field bees in what direction and how far away a new source of nectar may be found. Sometimes this bee dance is accomplished in as little time as fifteen seconds if the source of the nectar is very close, or as long as two minutes

Feeding A Weak Bee

if it is farther away. So in most instances we see this dance more by chance than by long observation.

Ant Alarm

Ants in the hive are a menace to our bees. After dark one evening my father and I went out to check on our hives. As we were leaning on the windbreak fence listening to the hum of the bees at work on the landing boards fanning with their wings to draw moist warm air from the hives I suddenly heard one bee give a peculiarly distinctive distress sound. We term it the "help me" call as the bees only make this particular sound when they are being invaded by ants and need the immediate help of their keeper. It is a sound somewhere between a whimper and a moan and is usually produced by one bee, though on occasion we hear it from many bees simultaneously. I think it is a combination of voice and fluttering wings. If we hear three or more bees making this distress sound the hive is really in trouble. Upon hearing the call for help I immediately went to the hive giving the call and knelt by it to listen. At intervals of a minute the call was repeated. I procured a flashlight and my Kellogg's Ant Paste (which unfortunately can no longer be purchased) and true enough, ants were invading the hive by running into it around one corner of the entrance. I traced the line of ants down the stand to its base where I placed a few drops of poison. The ants began to take the paste within a few moments. To safeguard the bees I covered the poison with a chip leaving enough space beneath it for the ants to reach the poison but not enough for a curious bee to reach it. To relieve and comfort the moaning bees I brushed off all the ants I could find that were running around on the landing board or near the hive entrance. Bees appreciate our help and will crawl around one's fingers in apparent joy. By morning there were no ants to be seen anywhere around the hive. Avoid using poison dusts and sprays which sometimes kill all of the bees as well as the ants!

Counting Queens

Much has been written about young queens who talk to each other while still in their cells just before emerging. Different observers record these sounds by various names but to me they sound like a rather high-pitched cheep or peep. These sounds are most often heard just at or after sundown. Many times there is considerable variation in the pitch and intensity of the peeps and I can determine with some certainty how many afterswarms to expect after the original swarm leaves the hive. There is a common belief that the first queen to emerge stings and kills all of the other young queens not yet emerged from their cells and thus becomes the sole survivor and inevitable queen of the hive. But in actual practice I have rarely found this to be true. The old queen flies away with the original swarm and at intervals of two to five days other emerging queens leave the hive with ever decreasing numbers of bees as second, third, fourth, and even fifth afterswarms. I have seen large hives almost depopulated after a series of afterswarms. Our own bees in those hives that we allow to swarm, almost always swarm three times and stop. I like that because we get the increase we desire and our strong hives still have enough workers of all kinds to build up rapidly and continue to make excess honey for us as long as the honey flow lasts.

Knowing the Swarm Signal

If a hobby beekeeper has a chance to observe his bees for a short time every day, and knows what to look for, he can completely control swarming in all of his hives without ever opening or looking through the brood chamber. My father and I have not searched a brood chamber for queen cells for many years, and in the year of 1978 we had no swarms issue from any of our hives because all of them had young queens and we did not need replacement queens nor increase. Rarely do we find a hive of bees that we call "swarmers," that is, bees that swarm five or more times in a season. When we do find

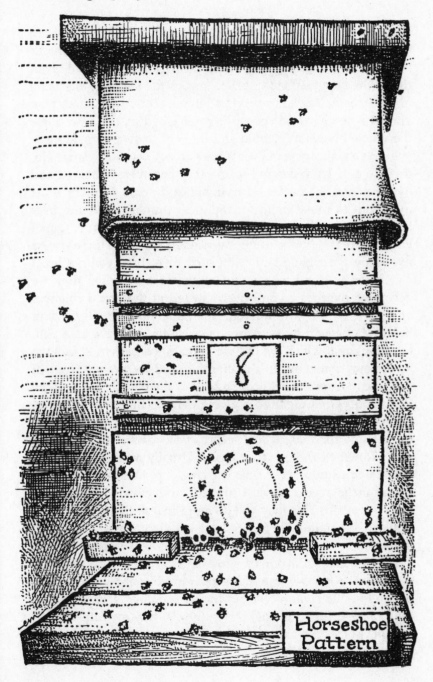

8

Horseshoe
Pattern

such a hive we requeen it as soon as possible by combining with it a new swarm from one of our best hives as given in our book *The Art & Adventure of Beekeeping*. Using the method given there, the newly added queen is always the victor in the battle of queens that occurs after combining. Thus in a short time after the newly added queen is established in her new home the whole hive is reoriented and the foolish swarming stops.

In the usual course of events bees do not want to swarm. No, they do not. They swarm because they become overheated and overcrowded in the brood chamber and some of them must move out to make room for the great number of young bees that are emerging every day during the spring buildup. If we give bees additional supers as needed and place these supers in exactly the right position on each hive, bees will not build swarm cells. Our bees give us ample warning when they are becoming overheated and overcrowded in the brood chamber and need our help to alleviate the situation. Three days before they begin to build swarm cells our bees give us a pantomime signal which we call the "horseshoe pattern." As soon as the sun comes up in the morning and warms the front of the hive most of the bees coming out of the hive fly away to the fields. But when the hive is giving definite thought to swarming, one, two, or three bees will crawl up the flat vertical front of the hive for four or five inches, veer toward the *right* and crawl down to the entrance again. Usually at about the same time an equal number of bees will crawl up the front of the hive a few inches, veer toward the *left*, and crawl down to the entrance again. This they will do more or less all day as long as the sun is shining. If fog rolls in or night comes they will immediately cease. The path the bees follow is very nearly the size and shape of a pony's shoe. It is distinctive and easy to recognize. When we first see this pattern our bees have not yet begun to build swarm cells but they are giving us warning that they plan to do so in the near future. When my father and I see this horseshoe pattern in progress we give serious thought as to what we should do. If we want increase or a new queen in that particular hive we do nothing, and we know that we can expect a swarm to emerge in about three weeks.

On the other hand, if the hive has a young queen, and we do not want increase, we know that we have four days in which to relieve the congestion. We must remove all of the supers from above the queen excluder and place an empty super just above it. This super may have either ten starter sheets to be drawn, or it may have ten fully drawn combs, or any combination thereof. Then we replace all of the supers we have taken off, placing the least filled super just above our newly added super and so on until the topmost one is the most nearly filled and sealed super. If after four days, or at the most five days, we have done nothing to ease the overcrowding the bees usually stop making the horseshoe pattern. That is why it is difficult for commercial beekeepers with outyards to observe the bees' swarming signal. Also, most of the large beekeepers use standard double depth brood chambers and bees do not always give the horseshoe pattern swarm signal when such large brood chambers are used. And even when they do there is not too much the beekeeper can do about it for the reason that in a double depth brood chamber the bees often jampack the upper one-third of the brood area with sealed honey, thus making a barrier that they will not willingly pass through to store honey in the upper supers, even though one places an empty super just above the queen excluder. They may go ahead and start swarm cells anyway. That is why for the first three years after hiving a new swarm we confine the queen to one standard full depth brood chamber, add a queen excluder, and place supers above it as needed. We never at any time use a brood chamber larger than one full depth brood box plus one medium depth super because we want to be able to control the congestion in the area of the hive just above the cells where the queen is actually laying eggs. This is the key to swarm control.

The year 1977 was another very dry year and during the swarm season we had to remove as many as eight to ten supers from our strongest hives as many as four times so as to be able to place an empty super just above the queen excluder. Our bees gave us the horseshoe pattern over and over again, but by paying heed to it I managed to keep all of them from swarming and we harvested a goodly amount of honey in spite of the drought as our bees were kept working at their maximum efficiency all season long.

The Sound of a Busy Queen

Our beehives emit a strumming sound during the spring buildup somewhat like the gentle picking on the strings of a guitar. As the season advances and the honey flow ends the strumming becomes less pronounced until about August 1 it ceases altogether until the next spring. This peculiar sound has always interested me. Others have heard it too and have asked me what it means. Until this summer I thought it was due either to bees squeezing through the queen excluder on their way to upper or lower parts of the hive, or to the vibrating wing tips of bees touching the queen excluder as the bees fanned with their wings to circulate air through the hive. But this year our greatest hive, No. 5, twelve medium depth supers plus a full depth brood chamber high, had no queen excluder for I removed it on April 1 so as to conduct another experiment. And I still heard the clear strumming sound I heard in all of our other large hives.

Most of the time I must press an ear tightly against the side of a hive to hear this sound, and it is always most audible in the area of the queen excluder. And it is still in the area where the queen excluder would normally be placed, that is, just above the brood chamber, that the sound is most pronounced, but obviously it is not due to the excluder itself. So I have come to the conclusion that the strumming sound is produced in close proximity to the queen wherever she may be, but by what means I do not know. In a large hive without a queen excluder advance knowledge of the whereabouts of a free ranging queen is of great help when servicing the hive, especially for those of us who do not believe in using Bee Go to drive bees down out of the supers. About nine o'clock in the evening previous to the day I want to work with our No. 5 hive I listen in on every super, mark the one with the loudest strumming, and the next day find the queen in the super that I have marked. Some have asked me why I do not use a stethoscope when listening to the bees in their hives. I have a stethoscope but for some reason not yet known to me I can understand the bees' language far better when pressing my left ear tightly against the side of a hive, listening in various places, than I can when using a stethoscope. I wish that some of you who read this account would also experiment along this line for it could help all of us to produce more honey. Theoretically, a hive without a queen excluder should produce more honey than one with an excluder for the simple reason that in a large hive multiplied thousands of bees must pass back and forth through the excluder almost every day. And each passage takes additional effort on the part of each bee. But in spite of this fact our greatest production every year has always been from a hive with a queen excluder. This fact led me to the conclusion that there must be something wrong with the way we handle our nonexcluder hives. And this thought prompted me to experiment with our No. 5 hive as mentioned above and continued in the next chapter.

12

Experiments In Progress

A Brood Chamber on Top

On APRIL 1, 1977 our No. 5 hive was already very strong and building up rapidly. It consisted of a full depth standard brood chamber, queen excluder, and seven medium depth supers above the queen excluder. I listened in on the bees and realized that on the next warm day I would have to remove all seven of the supers and place an eighth one down right over the queen excluder to alleviate the overcrowding immediately above the queen—or the bees would build swarm cells and fly away. I did not want increase from that hive so I planned to do what was necessary. But the next two days being cloudy and cool I had a chance to listen and think. Then the thought occurred to me, why not after removing the seven supers, also remove the queen excluder, detach the brood chamber with its queen and bees from its bottom board and temporarily set it aside, then place a well filled honey super on the bottom board, another lesser filled super above that one, and then replace all of the remaining supers on top of those, and last of all hoist the heavy brood chamber up on top of the stack, and work the hive for the balance of the season without a queen excluder?

The more I thought about it the more the idea intrigued me—so I tried it. Apparently the bees loved the idea too for they began to work more furiously than ever and in two weeks gave evidence that they needed more room. So I slipped off the heavy brood chamber and set it aside on a four foot high light-weight portable scaffold arrangement that my father had made for me, and placed two additional supers on the stack just below the brood chamber which I again set on top of the hive. The bees liked this too, and shortly after I had to add another tenth super below the brood chamber, and that is where the brood chamber has been all season, ten supers up. So far in 1977 I have removed three nicely filled supers from below the brood chamber and replaced them with supers to be drawn or filled.

One day during this episode my friend Joe came by. "You're wild!" he exclaimed. "Whoever heard of putting a brood chamber away up in the air like that?"

"But believe it or not—it works—and the bees love the arrangement," I answered.

"They seem to," he admitted after he had watched the great number of bees coming and going from the extended landing board. "But would they carry honey up higher if you placed a couple of supers away up on top above the brood chamber?"

"Oh, I don't think so, but now that you've mentioned it, let's try it and see. Sounds interesting."

So we did, and a month later when we looked in on the bees they had beautifully filled the topmost super with honey. The next one down, which was just above the brood chamber, was filled half with honey and half with brood.

"Why don't you go even higher?" Joe laughed as he spoke.

"I can't!" I answered. "I'm about to break my back now lifting these heavy supers back and forth—and my neck too—" as my improvised scaffold of wooden boxes gave a sudden lurch that threatened to dump me off as I clung to a super filled with partially drawn bee covered frames I was trying to replace on top of the hive.

"You'd better quit while you're winning," Joe agreed.

And that is where the matter stood as of July 21, 1977. Seventy-eight pounds of honey taken off with much more to be removed after it had thoroughly ripened. We took off the last of the honey about October 1, and No. 5 became our second best producing hive for the year, 153 pounds, a close rival to our No. 4 hive with a production of 155 pounds.

I do not know if I will try this wild experiment again. One thing is certain, a brood chamber filled with bees, brood, honey, and pollen is very heavy and I am not as young and strong as I used to be; but the lure of an unresolved experiment does beckon with ghostly power.

Grounding a Beehive

On May 2, 1975 we hived a fine swarm of bees into a brand new hive. The weather was favorable, the honey flow abundant. To our surprise and delight the bees drew out all ten frames in the brood chamber in only nine days, a record in our experience. I suggested to my father that this would be a good hive to use to try another experiment, the details of which had been running through my mind for some weeks. The idea I had in mind was to ground a beehive by running a length of copper wire through the wooden side of the queen excluder rim and then twisting it tightly around the cross wires of the excluder and grounding it to a rod driven into the ground.

"Why in the world would you want to ground a beehive?" my father asked.

"Because it may be that bees become charged with static electricity as I do. You know how much it helps and rests me to lie down on the ground for a few minutes. I can just feel the static electricity draining away."

"I've tried that too," he said, "and all I could feel was the hard bumps on the ground!"

"Well maybe so," I told him, "but bees won't have to lie on

the ground to be helped. All they'll have to do is continue to squeeze through the queen excluder and, hopefully, lose their electrical charge as their bodies rub against the excluder wires."

"Try it if you want to," my father agreed. "It might work."

So I quickly procured a new excluder, attached a few feet of clean copper wire, and placed the excluder on our new hive which we called No. 7. Above the excluder I placed the first super. Then I tried to drive a long slim rod into the ground but found that the ground was too dry and hard to insert the rod. So all I could do was curl the wire around into a circle or two like a ram's horn and let it dangle like an aerial. To our great surprise the bees drew that first super and had it well filled with nectar in just four days. We added another super and then others, some with drawn combs from other hives. At the end of the season our new hive had made an excess of eighty-five pounds of honey—a new record for us by a good thirty pounds. So we counted our "aerial hive" as we came to call it, a resounding success and spoke of it to many people.

Aerial and Grounded Hives

On January 30, 1976 our No. 7 hive was again very strong and ready for a super so we placed our aerial queen excluder and this time really grounded it, doing everything right, such as scraping the rod and also the copper wire so that it would make a positive contact. The results were spectacular from the beginning. The hive built up rapidly and continued to build up in spite of the dry and cool winter and spring. When we saw how well No. 7 was doing we placed wired queen excluders on other hives leaving the copper wires dangling as aerial hives so as to compare results with No. 7. Wired aerial No. 5 produced 222 pounds of honey in 1976, wired aerial No. 6 produced 224 pounds, wired aerial No. 8 had an excess of 203 pounds, and our wired and grounded No. 7 hive produced 310 pounds of excess honey, again for the second time breaking the old world's record of 300 pounds, but not our own 1974 record of 404 pounds. We did not wire our world's winning hive No. 4 because we wanted to use it as a control hive, and it produced 264 pounds. And though No. 4 always seemed to look as if it were our strongest hive, yet wired and grounded No. 7 produced a substantially greater total amount of wild flower honey by season's end.

We had one failure. We grounded our No. 4½ hive but I wired the excluder in an improper manner, as I realized after we had placed the excluder on the hive. But the weather was too cold to make any change so we left it in place throughout the season to see the final result. No. 4½ produced only 157 pounds, possibly more than it would have without the wire but not enough so that we could be sure. Except for our own No. 4 hive none of the unwired hives under my care, nor any others in the whole of California that have come to my attention, produced during the 1976 season more than 150 pounds of wild flower honey. And a great many hives did not produce nearly that much—some none at all. One thing is certain, we must carefully scrape all surfaces on copper wire, queen excluder, and rod, clean and bright. This we failed to do adequately on our No. 4½ hive in 1976. All copper wire seems to be covered with a very thin chemical coating that acts as an insulator to faint electrical current. We are continuing our experiments with aerial and grounded hives.

Experimenting with Color

In recent months I have begun new color experiments with our bees. About May 1, 1977 our No. 4 hive began to be somewhat cross in spite of our gunnysack wave cloth which continually swung around in the wind. The old sack had become faded and whitened with age. I began to wonder if a bright color such as green would help to tame the bees especially after I finished reading a little book entitled *Light and Color*, by W. J. Colville. So I fastened a two foot square piece of bright green cloth to the wave cloth rod arm in place of the old burlap. The results were immediate and spectacular. From the moment I put it up no more bees buzzed us or our visitors even though this hive is only ten feet from our back door. Since then I have put up brilliant blue, pink, and white wave clothes by other hives and they too are effective. But the bees did not seem to appreciate a large bright red piece of cloth that I tied on the front of one hive. They did not attempt to sting me when I went near to watch but they buzzed around the entrance in a rather frustrated manner so after a few days I removed the cloth. I tried it on another hive and they too seemed to find the red cloth detrimental so I have discarded that experiment.

Then one day Greg Smart, also a beekeeper, brought me some three-by-sixteen inch pieces of colored plastic that look much like glass.

"How about trying these colors on your bees?" he asked.

"Thank you. I'll be glad to try them. You've brought me red, blue, green, yellow, and amber. These ought to do something for my bees when the sun shines through them."

"Maybe more than you expect," he laughed.

"That could be too," I agreed.

I constructed a sixteen-by-twenty-four inch wooden frame with rabbeted slots so that I could place several pieces of colored plastic in it and slide them into various positions. Then I placed the frame in a horizontal position two feet above the entrance to our No. 4 beehive so that the sun could shine down through the pieces of plastic onto the bees coming and going at the entrance. First I tried the green plastic. The bees seemed to love it and walked around in its pretty light. I added yellow and amber, readjusting the green plastic so that the sun would first shine through it and automatically shift to the yellow and amber as the sun rose higher in the sky. Then I was called away and had to discontinue my observations for an hour. When I returned the sun had risen so much that it was shining through the amber pieces of plastic—and my bees did not like it! Half a gallon of them were climbing and milling about the hive entrance. In a moment several of them began to buzz me in an astonishing manner. I stepped back—and then ran for a bee veil—the first time this year that I have really needed one. I immediately removed my colored glass frame and the bees began to settle down. In half an hour they were completely quiet and were going on with their work as usual. Whether it was the amber color alone that so infuriated my bees or the addition of or combination with some other factor that so angered my bees I do not yet know. And my father will not let me continue the experiment at this time because one of our many visitors might be attacked. He is right too. As beekeepers we must take every precaution available to protect those who come to see us. There is still so much that none of us know about honeybees.

Experimenting with Hives

What the beekeeping industry needs is a canny inventor who can build us a beehive with a permanent rigid compartmented framework constructed in such a way that it will receive standard Langstroth hive components so that we can slip supers and frames in and out of it as needed to service our hives. The greatest difficulty, I think, would be how to overcome the bees' great propensity to stick everything together with propolis. When they get things glued together it is difficult to get them apart, to say the least. But there must be a way—if someone could just think of it. And it would have to be a method of which the bees highly approve so as to encourage them to enjoy life to the full, and thereby make more honey for us, naturally.

Love Those Honeybees

Honeybee Quiz No. 7
MULTIPLE CHOICE. Even though my beloved bees will not sting you if you miss more than 9 of these brainteasers, think positively and you will do much better. Consider what would be the logical answer before you make your choice, then mark every one. A score of 9 is average, 14 is excellent, and any score above that makes you a member of the prestigious Bee Fever Fellowship. *Answers follow the quiz.*

Experiments In Progress

1. We often find one of the following living as a colony in a paper-like nest in the ground: Hornets _____ Honeybees _____ Yellow jackets _____ Wasps _____ .

2. If honeybees living in a hollow tree succumb to disease the wax moths will: Seal the entrance hole so no other bees can enter _____ Enter the tree themselves to lay their eggs and so ultimately destroy all of the infected combs _____ Purify the combs with their saliva _____ Remove the honeycombs from the tree _____ .

3. A honeybee that falls into deep water will: Swim out in a short time _____ Drown immediately _____ Drown after an extended stay in the water _____ Fly up out of the water _____ .

4. If a hive is invaded by ants the bees will: Sting the ants _____ Bite them to death _____ Fan with their wings and try to blow the ants out of their hive _____ Live in harmony with the ants _____ .

5. Honeybees have been known to live and thrive in three of the following locations. The exception is: Wall of a house _____ In a nest under water _____ In a chimney _____ In a dry abandoned culvert _____ .

6. A swarming cluster of bees is never found: On an island in a small lake _____ Near the entrance to a hollow tree _____ Above the snow line _____ Under the eaves of a house _____ .

7. Which animal loves to eat bees: Hog _____ Skunk _____ Goat _____ Squirrel _____ .

8. The most advantageous height of a beehive above the ground depends upon all but one of these factors: Possibility of flooding _____ Ease of loading and unloading the hives _____ Rarity of the air _____ Presence of pedestrian or vehicular traffic _____ .

9. When is the time to put on the first super: After the bees have swarmed once _____ Swarmed twice _____ At the end of the honey flow _____ Just before the bees think of swarming the first time _____ .

10. If a mouse enters a beehive and dies there the bees: Encase the body in propolis to minimize decomposition _____ Gnaw the body to bits and remove it_____ Remove the flesh but leave the bones_____ Drag the body out of the hive _____ .

11. A honey extractor removes honey from the combs by: The use of heat_____ Crushing up the combs_____ Utilizing centrifugal force_____ Sucking honey from the cells like a vacuum cleaner_____ .

12. An electric uncapping knife has: A built-in light all along the cutting edge to aid the operator in watching his work _____ A heated cutting edge to aid in shearing off the cell caps_____ A single edge blade_____ A saw-toothed edge _____ .

13. Honeybees: Enjoy to sting _____ Prefer not to sting _____ Refuse to sting_____ Almost always die if they do sting_____ .

14. No man can: Handle honeybees without veil and gloves _____ Force bees to collect excess salable honey _____ Locate the queen among the thousands of bees in a hive _____ Estimate the distance a bee flies to its source of nectar _____ .

15. Honey in the comb is always: Dark colored_____ Light colored _____ Dependent for its coloring upon the source of the nectar _____ Without color_____ .

16. Beehives the world over are always: Built of wood_____ Built of available materials_____ Built rectangular in shape _____ Built of very heavy material to keep from blowing away in the wind_____ .

17. Bee sting venom is sometimes used in the treatment of: Hiccups _____ Snoring_____ New Year's Eve hangover _____ Arthritis _____ .

18. Honey is easily removed from clothing by washing in: Oil ___ Water____ Alcohol____ Cleaning solvent____ .

19. Basically, bees sting because: People are afraid of them _____ They become frightened or offended _____ They demand freedom of activity without human or animal in-

terference _____ We remove some of their excess honey
_____ .

20. A heaping teaspoonful of honey eaten a half hour before
each meal is an aid in: Growing hair on one's bald head
_____ Relieving an itchy nose _____ Diminishing the
growth of beard _____ Healing peptic ulcers_____ .

Answers:

1. Yellow jackets often live in the ground.
2. Wax moths enter a dead hive to lay their eggs and so
   ultimately destroy all of the infected combs.
3. Honeybees drown after an extended stay in the
   water.
4. Fan with their wings and try to blow the ants out of
   their hive.
5. In a nest under water.
6. Above the snow line.
7. Skunks love to eat bees.
8. Rarity of the air.
9. Just before the bees think of swarming the first time
   add a super.
10. Encase the body in propolis to minimize decomposition.
11. Honey extractors work by utilizing centrifugal force.
12. A heated cutting edge to aid in shearing off the cell
    caps.
13. There are two correct answers. Honeybees prefer not
    to sting, and almost always die if they do sting.
14. Force bees to collect excess salable honey.
15. Honey is dependent for its coloring upon the source
    of the nectar.
16. Built of available materials.
17. Arthritis is sometimes treated with bee venom.
18. Water washes away honey.
19. Bees become frightened or offended and then they
    sting.
20. Honey can be an aid in healing peptic ulcers.

Appendix A

Answers To Beekeeper's Questions

1. Question: *Is it easy to get started with bees?*

Answer: Yes, compared to getting started in other major hobbies, or small businesses. Books on every phase of beekeeping are available in many libraries and bookstores and almost everywhere one can find congenial beekeepers who will take the time to help a beginner with some of his initial problems, the first one of which is usually getting that first swarm hived and taken home. There are beekeepers organizations in many places that welcome visitors and new members. The local agricultural office can often give one information about these groups and where they meet.

As to the expense, it now costs from fifty to seventy dollars to buy a new hive with a rousing swarm of bees newly hived. Add to this the cost of a beginner's outfit, additional beehive supers and equipment that will be needed before the second bee season begins, an extractor, electric uncapping knife, extractor room and miscellaneous other expenses and the cost comes to around $500. Many who have tools and the know-how to build much of their equipment can get started for less money, but even they almost invariably begin to increase their number of hives so that they too soon need $500 or even more. My friend Lee is a good example. Two years ago he started with bees.

"I want just two hives," he told me. "Just two and no more."

"How many hives do you have now?" I asked him a few weeks ago when he came over for a visit.

"Thirteen! And now it's always buy more equipment to hive more bees—or add supers to the hives I already own. Then buy hives to catch the swarms that come off. It's a never ending circle! And it takes all of my spare cash!"

"Of course it does, if one continually expands his operation," I told him. "Why don't you sell some of your hives?"

"Well, I'm thinking about it. But they're all such good bees!"

And now there is Greg Smart, another beekeeper who often stops to see me. Last year he had two hives—now he has six. Both of these young men, and there are young women just like them, have delightfully severe cases of bee fever and are going to make real successes of beekeeping. Lee already had much honey to sell last season and he will have more this season. It is sometimes easier to get started with bees than it is to keep up with them.

2. *Q: How do you ventilate your hives in winter? I have been told that bees will die out from dampness, lack of ventilation rather than cold.*

A: The above question and comment came to me from Alice Howell of Resort of the Mountains, Morton, Washington. Bees dying from dampness due to lack of ventilation has not been our experience and we kept bees for many years in the Willamette Valley of Oregon as well as various locations here in California. But we have never had the care and observation of bees in the area of Morton, Washington, a region with a higher elevation than anywhere we have ever kept bees. In Oregon we had our bees on stands at least two feet high to keep them above all but the heaviest snowfalls. When it snowed (very wet snow) we watched our beehives as they were always near our house for the winter, and we were careful to keep a hole poked through the snow to each entrance so that the bees could get air. We always had our hives facing southeast so that any sun at all would help dry out the hives. And of course our hives were always placed so that the sun could reach them all day long if there was any sun. We never placed our hives under trees of any kind. To protect them from the rain we used roof covers, metal or boards, two and one-half feet by three feet in size, weighted down with at least forty pounds of rocks. If the wind was blowing a gale we also placed sloping shield boards on the windward side of each hive to keep it dry from the blowing wind and rain. 🐝 Lastly we always were careful to see that the back end of each hive was just one inch higher than the front end so that any rain blowing into the hive could drain out by gravity. These precautions really paid off for us. We lost some hives during each winter, of course, but not due to dampness alone, rather to a combination of factors like a failing queen or insufficient stores left on the hive for the winter. 🐝 For many years now we have made it a rule to leave on each hive at least ten pounds of honey more than we think the bees can possibly use for the winter, and that has paid off handsomely. We all have a tendency to rob our bees and that is sheer folly. We never, ever, take honey from the brood chamber, and we almost always leave on a medium depth super quite well filled with honey for additional stores. Once in awhile our bees find a late nectar flow and are able to pack their brood chamber sufficiently to carry them easily through the winter, but not often, and we take no unnecessary chances. We leave plenty of honey as it can always be taken off early the next season.

A friend of mine used to live in Minnesota where his family had several thousand beehives. There they used double depth brood chambers (we never do here) and to prevent the bees from suffocating for lack of air during a severe snow storm or prolonged severe cold spell when the regular entrance would, or could, become plugged by dead bees, they bored a three-quarter inch diameter hole in the front end board of the upper brood chamber. This airhole allowed foul moist air to escape. Some warm air escaped too but even so their hives wintered well as far as ventilation was concerned.

3. Q: *Do you always use queen excluders?*

A: Not always. Queen excluders are a definite help to the beekeeper and as a rule we use them. Those hives that we want to produce honey exclusively so as to have honey to sell to our customers we run with queen excluders as it makes it easier and faster for us to service such hives. But we plan to have at least one hive, for experimental purposes, that has no queen excluder. Theoretically a hive without a queen excluder should produce more honey than one with a queen excluder because it takes effort on the part of the bees to squeeze up through the excluder or down, time and time again every day. Yet through the years our greatest production has always been from a hive with a queen excluder. Why this should be true we do not know, but we are still trying to find out.

Sometimes we find bees that are cross hybrids of various combinations, and some of them are too large to pass through a queen excluder. Then we are compelled to remove the excluder. If after we place a queen excluder and first super in the spring of the year, and ten days or two weeks later find few if any bees up in the super, we suspect that we have oversized bees and remove the queen excluder. As a rule one can see, just by looking at the bees, that they are oversize. If after careful scrutiny they appear to be normal size we wait another week or two before removing the excluder for the reason that the bees may not have had time to build up enough in numbers to find it necessary to go up into the super because of the need for more space. When they need the space they will go up unless they are oversize.

Most of the time when keeping bees in Oregon we had hives that consisted of a full depth brood box plus a medium depth super comprising the brood chamber. This provided the bees ample space for brood rearing and also for winter stores. We added a queen excluder and supers in the spring as needed and removed them in the fall, usually in September. Down here in Santa Cruz, California, we usually use a single full depth brood box for the brood chamber and leave on a queen excluder and medium depth super all winter. We are experimenting with various combinations and often also use two medium depth supers for the brood chamber (we made our world's record with these) and for the winter we leave on the queen excluder and above it a well filled medium depth super. Then in the spring we raise the super and slip a second one beneath it. The bees like this arrangement and go right to work on the newly added super.

4. *Q: When you have a hive composed of a double brood chamber how do you change it over into a single depth brood chamber?*

A: I like to make the exchange and consolidation of frames during the months of September and October. During these months in our area the days and nights are still quite warm and the bee population is much diminished due to the approach of winter. On a warm windless day I pry up the hive cover and force a few small puffs of smoke under the partially lifted hive cover just before removing it. Then I set the cover aside and carefully scrutinize the bees on top of the frames. If they are clustered thickly together over most of the top it often signifies that the queen is living in the upper half of the hive. So I immediately break apart the two parts of the hive and set the topmost part aside on saw horses set out for that purpose. Then I examine the top of the frames on the lower part of the brood chamber. If I find noticeably fewer bees on the top bars of those frames I consider that I am right in assuming that the queen and brood are for the most part in the upper part. I immediately loosen the bottom part of the hive from its bottom board, set it aside, and install what had originally been the topmost part of the hive down on the bottom board.

If the day is far spent or the weather begins to turn cold I place a queen excluder on the lower part of the hive and place the former lower brood chamber above it as a full depth storage super and wait until the next warm day to examine these super frames one by one to see if the queen is actually down below where she should be. On a suitable day I remove the frames one at a time being careful to part the first frame slightly from the frame next to it and force a few puffs of smoke between them to drive the queen down to the queen excluder if she should by chance be yet in the super. Then I remove the frame, brush the few remaining bees onto the remaining frames, then set it aside. I remove each frame in its turn, and then the empty box itself. By this time most of the bees have gone down through the queen excluder and one can usually see at a glance if the queen is on the excluder. If she is, I very carefully slide the excluder to one side and ease the queen over the edge and down onto the frames below. I like to have a piece of cardboard handy to place over the hive at this time to make it somewhat dark above the queen so that she will not so easily think of taking wing and flying away. Great care must be exercised at this point if one sees the queen. But as a rule I do not see the queen because she has already been living below the excluder.

Now when I look at the frames I usually find four or five that are still in quite good condition which may even have some eggs or sealed brood. I set these aside. Next I remove the queen excluder and carefully pry out two frames on each side of the brood chamber. Rarely during the early fall months will one find brood in these outer frames. If they are in excellent condition and quite well filled with honey I replace them in the same position as they were before I removed them. But if they are in bad condition or practically empty of stores I replace them with the four best frames that I had set aside from the storage super. That done, I replace the queen excluder and storage super with all of its frames, both good and bad, and replace the hive cover.

If some of these frames above the queen excluder contain eggs or larvae or sealed brood, no matter, for as they emerge they will be able to pass down through the excluder, all except the drones. If there are many occupied drone cells I bore a three-quarter inch diameter hole in the front of the storage super and every four or five days pull the wooden closure plug to release the emerged drones. They are quick to detect an escape hole and in fifteen minutes I can replace the plug.

Now on occasion one will find almost an equal number of bees on the top bars of the frames above and the frames below. Then one can only guess where the queen is, or if the weather is sufficiently favorable one can remove the frames and look for the queen. But with my poor eyes I never go to that much trouble, for all things being equal, I go right to work to remove the lower part of the brood chamber from its bottom board and immediately replace it with what had been the upper half for the reason that usually the upper half is in better condition than the original lower half. But in all cases the beekeeper should do what seems best for that particular hive.

I have answered the question asked above but that does not mean that I always advocate making the change. If one has but one or two hives I would not take the risk of changing them over, but would let them swarm naturally the next spring and catch the swarms as they emerge until I had several more hives safely housed in single depth brood chambers. Also, one can usually obtain considerable honey from a double depth brood chamber hive—but for absolute maximum honey production I always do better with a smaller brood chamber.

5. *Q: How soon after servicing your hives, that is, after adding an empty super or taking off a filled one, do you again go among your bees?*

A: I try to keep completely away from them for at least three hours, overnight would be better. When bees are again taking up their housework after our high-handed interruption they are inclined to resent us if we try to approach them too closely a second time. I find that more people are stung after they have finished servicing their hives than they are while working with them. If I have dropped my hive tool or bee brush, or have forgotten to do something that should have been done I let it wait until evening or the next day before trying to retrieve or rectify it.

6. *Q: You say that you smoke yourself more than you do your bees. Do you also smoke your equipment such as veil and gloves before working with your bees?*

A: Yes indeed. I smoke all of my equipment thoroughly before again using it in order to eliminate the odor of bee venom as well as cover up my own human scent. Washing protective suits, if one uses them, is essential also if one wishes to keep his bees very gentle.

7. *Q: What about ants? How soon after finding a colony of ants invading a hive should one put out ant poison or take other remedial action?*

A: Immediately. Even a relatively few ants crawling all around the entrance of a hive tends to demoralize the bees. The poor bees have no positive defense against ants and they cry out for help from their keeper. Kellogg's Ant Paste seems to be no longer available so in its place we use Grant's Ant Stakes or other products of a like nature. But beware, bees as well as ants like these preparations and can be poisoned by them. We must be sure to put some protective covering over the drop or smear of poision, such as a small wooden box or cardboard carton. In some cases it is easier to poison the ants by opening the stake and removing some of the jelly-like paste and smearing a dab of it on the leg of the stand in the ants' runway.

A truly better preparation is Tanglefoot. This comes in a tube like toothpaste and can be spread around in a narrow band on the legs of the stand. Ants will not cross it until enough dust has blown on it to counteract its stickiness. When this happens it must be reapplied. Tanglefoot is nontoxic and I like that factor.

During the past two years several of those who have read my first book on beekeeping have sent me ideas on how to rid hives of ants. Some of these ideas have merit but they are rather hard to apply to established stands, especially some of mine where the stand posts are driven into the ground. Just now as of June 11, 1977 I received a most interesting idea from Mr. J. W. Binsack of Staten Island, New York. He writes, "I'm writing this letter in order to give you a helpful idea that I found will keep ants from climbing our two dwarf apple trees and it should also prevent them from entering your beehives. It's a cone of wire screen, sixteen by eighteen mesh which is small enough to keep the smallest ants from passing through the screen. The cone of wire is tied around the trunk and flared out on the bottom, extending down about eight inches. It works this way. The ants climbing the trunk come to the screen and try to continue up. They sometimes circle around on the inside of the wire but always try to go up and not down and over the lower edge of the screen. They're programmed to go up and always try to do this.

"To protect your bees could you fasten an eight inch wide wire screen around the platform that your hives are resting upon? You could fasten the screen to narrow strips of wood, the length and width of the platform, with staples. Then these narrow strips with the screen fastened to them can be screwed to the edge of the platform on all sides. There should be no cracks or openings. It is also best to angle the screen slightly out rather than to leave it straight down. I don't know if California ants are more intelligent than Staten Island ants but this works 100% for me as they soon get the message and don't try to climb the trees after a short time. Best wishes for your continued success."

And my heartfelt thanks to Mr. Binsack, Kathy West, and others who have been good enough to share their ideas with me.

8. *Q: Do you always paint your hives white or aluminum color?*

A: No. We obtain better maximum production if we leave the wood its natural color of red, brown or yellow. Many times we do not even paint or protect the wood in any way. This is especially true of our supers as most of them are only out on the hives about four months in the year and most of that time they are covered with plastic to conserve heat in the hives so that our bees can work more efficiently. However, when we were living in hot dry areas we had to paint our hives white or aluminum color to help dispel the direct rays of the sun and also to give some measure of protection to the wood. Usually beehives are figured on a depreciation rate of ten years. We calculate ours on a six year basis. But since we get so much more honey we can well afford to do so.

9. *Q: What do you do about mold on frames and wax of a used hive?*

A: With a hive tool I scrape from the frames, walls, and bottom board as much mold and gook as I can and then wash with a jet of cold water from a hose. I reassemble the brood chamber and hive a new swarm into it. The bees quickly clean up any mold remaining on the frames and drawn wax.

10. *Q: Must one always see the queen enter the hive when hiving a new swarm?*

A: No. I rarely see the queen entering with the other thousands of bees for the simple reason that I do not take time to watch for her. I know that if the bees stream into the hive for ten minutes or so—and stay in—she is among them for they will seldom go in and stay in without her. But if after ten minutes or so I notice the bees begin to pour out of the hive again I immediately begin to search carefully on the ground, shrubs, and everything else in the immediate area to look for the queen. Usually I find her under a tiny cluster of a few dozen bees either on the ground or somewhere close by. Then I gently urge her to enter the hive and the bees all turn around again and go in after her.

A few days ago a friend of mine, in the process of hiving a swarm, saw the bees turn and start out again. Instantly he glanced at the ground near him and saw the queen and three escort bees only inches away from his foot. She had not gone in with the swarm and as he stepped back to watch he had almost stepped on her. He urged her onto a leaf and carried her the few feet to the hive entrance. She hurried in and the other bees, scenting her presence, turned around and hurried in again too.

11. *Q: In the event that you find burr comb, that is, drawn and filled cells wedged in between the bottom bars of one super and the top bars of the super below it, do you cut off this burr comb or do you leave it for the bees to refill?*

A: I cut it off. I do just skim off the caps of the cells on both sides of the frame proper to make it as easy for the bees to refill with honey as possible but I always remove all extra propolis or wax burr comb from both the top bar and also the bottom bar or rails. If one does not, one is too apt to crush bees between these frame parts when replacing the supers back on the hives for the bees to refill. Such crushed bees, in their agony, scream and may cause the guards to really sting.

12. *Q: Is is possible for people to hear the cries of bees as they are being crushed?*

A: Yes, if they have exceptionally good hearing and ears that can detect very high sounds. Until the last year or two I could always hear the distressed cry of a bee if I was in some way squeezing her too hard, and of course I made an instantaneous effort to ease the pressure. At the present time I know two young lady beekeepers who can hear the voices of honeybees. Every beekeeper should attempt to attune his ears to the cries and sounds of his hard working little servants.

Screened
air vents

Swarm Catching
Can

13. *Q: Do you ever find a new swarm and carry it home in a sack or cardboard box if you have no standard hive prepared and ready?*

A: Yes. In an emergency honeybees may be temporarily hived into almost anything that will keep them confined and at the same time allow them to get plenty of fresh air. A loosely woven gunny sack may be used, but one would never use a large plastic bag as the bees would suffocate if there are no holes in it and the plastic bags that I have seen that do have holes, have such large ones that the bees can escape. A cardboard box such as would hold a dozen fruit jars may be used for smaller swarms and a correspondingly larger box for larger swarms. In one end of such a box we cut a rectangular slot or hole about one inch high and twelve inches long to allow the bees entry and also air. When they are almost all in the box we tie a piece of window screen over the hole and take them home. However, it is much better to be prepared in advance with a beehive all ready and waiting so that we can hive the bees directly into it as soon as we get a call asking us to hive a swarm. A new swarm has only a basic amount of strength which varies with the number of bees in the swarm. We want the bees to conserve and use all of their strength in building and drawing out the wax of the brood combs and storing honey and pollen as soon as possible so that the queen can begin laying eggs, for no new strength will be added to the hive until the new brood emerges. Rehiving bees for whatever reason is wasteful of the bees' strength and resources. Be prepared— have your permanent hive ready in advance.

14. *Q: Sometimes when I take off supers of honey I find bees in some of the empty cells with just their tails sticking out. What are they doing?*

A: Sleeping. Especially in large hives having five or more medium depth supers above the queen excluder we often find numbers of bees that are asleep in any available empty cell. Bees can sleep and rest in any position, vertical with head up or head down, or sideways, but they prefer to crawl into a cell head first and go to sleep lying on their stomachs, as we would think of it. When taking off honey I handle these sleeping bees with care and consideration. After removing a frame of honey I gently sweep my bee brush over the bees whose tails are protruding from the cells. This usually arouses them and they begin to back out of their cells. But soundly sleeping bees may not awaken so I let them sleep. By the time my father has carried the frames to our honey extracting room they wake up and fly away. Rudely awakened bees tend to become cross and threaten to sting—and we cannot blame them for that.

Very large hives such as those with seven to eleven supers above the queen excluder require what my father and I call "bedroom space." These are medium depth supers with drawn combs that we place down about five supers from the top of the hive. If about two weeks later we find that the bees have begun to fill our bedroom super with honey we give them another in the same location as well as service the hive in whatever other way it may require. For maximum production during the height of the honey flow we must provide adequate rest and resting space for our eager little workers who often toil all day and all night to gather in the golden harvest while it is available. Our bees realize even more than we do that time is precious.

15. *Q: How can you get so much honey when bees have so many diseases and calamities and enemies? I'm almost bluffed out of bee-keeping before I start!*

A: Well, take heart! Compared to human ills and accidents, bees have almost nothing against them and everything going for them. And if we give them thoughtful and loving care and understanding they can not only survive but joyfully gather excess honey for us.

16. *Q: Do teenagers ever keep bees?*

A: Yes. Many young people, both boys and girls, begin keeping bees at an early age and we who are older should encourage them in this endeavor. Bees are fascinating little creatures that almost anyone can learn to keep, basically speaking, and they largely care for and feed themselves.

One of the youngest successful beekeepers of my acquaintance is Joel Govostes, who in 1976 was an eleven year old beekeeper of three hives which in 1975 produced 150 pounds of excess honey for him. After reading my first book he wrote me a beautiful fan letter. His card reads Joel Govostes Apiaries, 320 Salem St., Woburn, Massachusetts, 01801.

My friend Randy Anderson, age fourteen when he first wrote to me, telephoned me from his home in Williamsville, New York on May 31, 1977 that he really had some bees! He was bubbling over with joy—as well he might be—for he had been trying for a year to acquire a hive of bees. Earlier he had located a bee tree, only to have the owner of the tree give the bees to some experienced beekeepers who would not listen to his plea that he had found the bees first and he wanted a swarm so badly. The hard hearted characters took the bees anyway. And then he just missed out on a beautiful swarm—and so it went all the time for a year. Then he met a boy named Sho who was also interested in bees and together they built a hive and then ordered package bees from Florida. Cost what it might, he had to telephone me the good news that the bees had arrived safely. A recent letter states that his bees are really prospering with new brood emerging.

17. *Q: Why do you nail your full length handholds at the top edge on each end of your supers and use a correspondingly narrower one at each end near the bottom edge?*

A: For two reasons. When we build up our hives in the spring of the year seven to eleven medium depth supers above the queen excluder we often find it necessary to stand the supers on end for the simple reason that there is not enough room around us to set them all down flat. Built the way we do they will stand upright on end, bees and all, until we want to replace them on the hive. If we just nail one full length cleat across the center of the end, the super will tip over. The same holds true of only a top handhold and no bottom one of equal thickness. It takes both to assure that the honey-filled frames with their adhering bees will remain upright. The bees really become angry if a partially filled super tips over.

The second reason is that in the usual super the upper edge of the end board is rabbeted out to only half its thickness so as to leave a kind of ledge to receive the ends of the honey frames. This makes the top of each end board very weak at a point where there is often much stress due to the bees using propolis to glue the supers together. To break this seal we often have to insert our hive tool between the corners of the supers and exert enough force to break them apart. This sometimes damages the end pieces of the supers. When we use full length handholds and full length counters, much of our force is applied to these pieces, and if they split or break, we can replace them. Also we do not need to be so very careful to get the supers all lined up again when we replace them on the hive for the handholds give us additional leeway endways.

18. *Q: Do you ever use hives with four rabbeted fingerholds in each super such as the large manufacturers make?*

A: Not any more. Honey is heavy and all sizes of supers when filled with honey are heavy—too heavy for my fingertips to handle. And I notice that others also occasionally drop a super of honey and that makes for a real mess with some of the honeycombs broken out of the frames. Such combs must be painstakingly replaced and tied in again with cotton string before they can be extracted and that takes time. As a hobby beekeeper it definitely pays to save broken combs and have the bees repair them rather than have them draw new combs. But as a commercial beekeeper I would not bother as the time and expense involved would be too great.

Also in our cool Santa Cruz, California area we harvest more honey from each hive if we use full length handholds. Lumber is ⌐ so thin these days that the rabbeted fingerholds almost cut a hole through each of the side and end boards of every super. This results in a cold spot in the hive wherever one sees a fingerhold and the bees do not work these areas as readily as if all of the boards were full thickness. In very warm parts of the country I might still be using factory type equipment.

19. Q: *Will you please discuss plastic core super foundation?*

A: Special care should be exercised when using plastic core super foundation. There seems to be a common belief among beginning beekeepers who do not want to go to the expense of buying a honey extractor that they can, when the frames are drawn out and the honey in the combs sealed, simply scrape off right down to the plastic core, all of the drawn out wax comb with its cappings and honey. The idea is to later use a colander to let the honey drain away from the wax. One can do this, and the bees will again draw and rebuild the destroyed comb using the plastic core foundation as a starter provided one has not so thoroughly scraped the plastic center sheet as to completely remove the coating of beeswax. Visitors often bring me frames with plastic core foundation that the bees will not redraw, or will redraw only in part, and they want to know why the bees would not redraw the entire sheet. The reason is that bees will not ordinarily willingly draw on uncoated plastic even though the cell base pattern is imprinted in the plastic core. Bees want a coating of beeswax upon which to start their work. My advice is to crush the cells and remove that which comes off easily and return the remainder to the bees. They will quickly clean up any liquid honey remaining and then begin to rebuild the combs.

In these later years one can also buy all plastic frames. No wood at all is used in the construction of the frame and all parts, back bar, end pieces, bottom rails, and the starter sheet itself are molded as a unit. Bees like to work on these frames and draw out beautiful straight combs. These all plastic frames are designed for use in the new high speed extractors which do not require that the combs be uncapped. Uncapping takes time and labor and for the commercial beekeeper these new frames are excellent as he can afford the approximate $1500 initial expense of buying such an extractor. But for the backyard bee-keeper who has the usual low speed extractor these frames are difficult to handle and uncap. They are so slick and the plastic is so tough that if the uncapping knife touches the frame at all the knife stops and one has to back off and try the cut again. With wooden frames all we do is cut off a chip or sliver of wood and hardly know the difference as the hot knife skims through the caps.

20. *Q: How do you control drone rearing?*

A: People often tell me that they are hindered in their pro-duction of honey by having too many drones. Their hives seem to be alive with drones all spring and summer long. Yet when they observe our hives they see few, if any, drones. Years ago my father and I had the same problem so we began a series of experiments and observations to find the reason. We have come to some interesting conclusions and suggest that others also experiment along the lines of our observations.

As presently manufactured, brood chamber frames are self-spacing when pushed close together. The upper shoulders of the frames hold the wired foundation sheets exactly one and three-eighths inches apart. To hold the frames in this exact close proximity to each other when the hive is new we always posi-tion the ten brood frames while the hive is yet without bees, and pin them into their exact location with tiny nails or brads. Four brads, one at each corner of the hive next to the frames, are sufficient. One does not need to pin each frame individually. Then when we hive a swarm of bees we do not have to be so careful about jarring or tipping the hive as the frames stay in their exact position due to being pinned in place. If not firmly pinned the frames are sometimes shifted by the bees them-selves. On occasion our experiments showed that even little honeybees, when they make up their collective minds, can exert astonishing pressure, enough to move apart the center frames in the hive. They squeeze their bodies in between the frames and push until the frames move.

Long ago we learned that if any of the brood frame shoulders become separated as much as one-quarter inch from the one next to it, for whatever reason whether during hiving or at a later date, the bees will almost invariably utilize most of the additional space between the frames for drawing the larger drone cells—and when the queen comes to them she will lay drone eggs—and we have drones in great overabundance. But ever since we have been careful to pin our brood frames before we hive a swarm, and rarely remove the brood frames after the bees are in the hive, we have had no problem with excess drones. Every strong and healthy hive will have some drones. That is a fact of life with honeybees and we are always glad when we see the first drones around the entrance of a hive in early spring. Even with properly spaced frames and wired foundation sheets the bees will tear out some of the worker cells and replace them with the larger drone cells. This is good. But we can greatly minimize this tendency by adding additional supers on the hive as needed, or just a little sooner than they are needed, or so it always seems to me. It is better to be a few days early in adding a super than a few days late if we want to control drone rearing and swarming.

21. *Q: How do you control swarming? Do you look through the brood chamber every ten days to destroy the queen cells?*

A: We never look through the brood chamber to destroy queen cells and our bees never swarm unless we want them to do so for the purpose of requeening or increase. We have no problem with our bees continually building queen cells and swarming. An excess of drones and an excess of queen cells seems to be very closely related and we control both to a large extent by our careful positioning of the brood frames. We sometimes use a straightedge and draw a diagonal line across the backs of the frames as an aid in replacing the frames into their exact position. Such a line is especially helpful for those who must remove brood frames for examination or other reason. Exact replacement in proper order is absolutely essential for maximum honey production. An excess of space next to either or both sides of the brood chamber is no hinderance as we rarely see either drone cells or queen cells on the outside of a frame next to the hive wall.

🐝 Double depth brood chambers have a tendency to cause bees to swarm no matter how many supers or how much room the bees may have above the queen excluder. The reason, in our experience, is that few queens, not even our world winning queen, need that much space for egg laying. The result is that the queen lays eggs in the cells that she can cover and the bees fill the balance of the cells with honey and pollen, usually in the upper one-third of the double depth brood chamber. And they often fill this upper one-third of the brood chamber to its utmost capacity thus greatly restricting the circulation of air to the higher supers and also making it difficult for the worker bees and hive bees to pass up and down between the fully drawn and sealed combs below the queen excluder. This honey barrier results in overcrowding in the brood chamber as the season advances with the result that the bees begin to build swarm cells. In two weeks the primary swarm flies away and it may be followed by several secondary swarms, at intervals of a few days or even a week.

We cannot prevent the bees from building such a honey barrier but my father and I control it by using only a single full depth brood chamber on most of our hives. There are exceptions as previously stated in earlier chapters. With the queen confined to only one full depth brood box the bees make maximum use of that space for the queen's egg laying and store practically all of their pollen and honey in the super just above the queen excluder. This is exactly what we want them to do, for when it becomes jampacked with pollen, honey, and bees, we raise it up and slip an empty super under it for the bees to work in and fill. This they are glad to do for they have been relieved of the heat and congestion just above the brood nest. This dry season we have had to give our strongest hives an empty super just above the queen excluder four different times. It necessitated a lot of work taking off from seven to eleven supers so as to be able to place an empty one immediately above the queen excluder but it kept our bees from swarming. I did not want any swarms this year—and we had none. 🐝 Bees are not anxious to swarm. They swarm because they feel they must due to unbearably crowded conditions in the area of the brood chamber. When they have breathing space and elbow room they will gladly work high up in the stack of supers and produce a great amount of honey.

22. *Q: Do you think that the hobby or backyard beekeeper has a production advantage over the outyard beekeeper?*

A: Yes. In many instances he does. Backyard beekeepers who have the opportunity to observe their hives every day have a great advantage over beekeepers whose outyards are at a distance because the backyard beekeeper can continually observe his bees and do what is necessary to keep his bees busy and happy. Sometimes bees gain in numbers more rapidly than we could have foreseen so we add a super when it is needed. Again they build up more slowly and we wait a few days longer to add the super. Or the weather turns cold or rainy and we replace a restrictive entrance closure, temporarily, that we had already removed during a period of warm weather. The outyard beekeeper must of necessity make an educated guess as to what each hive requires for it may be some weeks before he services his hives again.

23. *Q: Do you ask parents to keep their children away from your beehives?*

A: No. We never ask the parents of children to keep their young folks away from our beehives. Many children find bees fascinating—and we want them to continue to cultivate this wonderful gift of love and life, for they are going to be the beekeepers of the future—and without more bees we will all go hungrier than we are now. As backyard beekeepers we must protect our bees and safeguard the children by building a three and a half or four foot high fence all around our hives. Almost any material available may be used for building the fence. If we use wooden pickets or boards we place them so that they have cracks about one-quarter inch wide between each picket to allow a small amount of wind to pass through. Such an enclosure, adequate for four hives, would be about twenty feet long and fifteen feet wide. To make it possible for little children to watch the bees we cut an opening in the boards or pickets and cover it with wire screen to make what amounts to an observation window. Children love such a window and can learn much by observing the bees at work as the season passes.

24. *Q: How do you know when a newly hived swarm has completely drawn the ten brood frames? You so heartily disapprove of opening a hive every day or two to see how the bees are progressing. Is there a better way?*

A: Yes. As you know there is great variation in the time it takes for a swarm of bees to draw out the wax on the ten foundation frames. One of our swarms last season had them all drawn out and largely filled with brood, pollen, and honey at the end of nine days—something of a record in my experience. Usually it takes a good-sized swarm from twenty-four to thirty days to do the job. During the initial drawing period I always keep the entrance restricted to about one-fourth its total size by using restrictive entrance cleats to help conserve the heat of the hive. To check on the progress of a hive, during the heat of the day I remove all cleats leaving the entrance wide open, or if it is a sunny but windy day I use a special entrance cleat that restricts the opening to three-eighths inch high by the full width of the opening. Then I watch the field bees as they come and go. Bees always like to go directly to and from their work in a hive. Within an hour they will begin to change their landing and leaving pattern and go directly through the entrance to or from the frame they are working on. If they have drawn and filled four frames they will utilize that part of the landing board that leads directly to those drawn frames. As they draw more frames they will use a correspondingly greater width of the entrance opening. When they use the complete width of the entrance it is a sign that they are probably ready for a first super. So I set out a prepared super in the sun to warm up and also a queen excluder, and then remove the hive cover for a quick visual check of the hive and, if the drawing is all completed, immediately add the queen excluder and super, and close the hive.

Another way to check to see if the frames are all drawn, if one must be away during the warm part of each day, is to wait until evening and then remove all of the entrance cleats leaving the entrance wide open. Then with a rather dim flashlight shine the beam into the entrance and under the frames. As a rule the bees will mass under the frames they have completely drawn and in which there is brood. The problem is to replace the removed entrance cleats without crushing some alert guard bees that sometimes come out in force. When the bees under the frames are almost across the width of the hive, open the hive and make a visual check on the first available warm afternoon. Do not wait too long or the bees may build swarm cells due to overcrowding. But never try to make the visual check at night as bees definitely resent being disturbed at night.

25. Q: *Is it necessary for all of the brood frames to be completely drawn before adding a queen excluder and super?*

A: Yes. For as a rule upon the addition of a super the bees immediately stop work drawing out the remaining frames in the brood chamber and begin work in the newly added super space provided, for the reason that bees prefer to work vertically rather than horizontally. We must compel them to finish drawing the last of the brood frames or the queen will be short of egg laying space for years to come, thus depriving the hive of its opportunity to produce honey at its maximum capacity.

Time To place first Supers

26. Q: *What happens if one waits too long to place the queen excluder and first super?*

A: In all probability the bees will quickly become overcrowded, build queen cells, and swarm. If one wants to increase his number of hives this is cause for rejoicing. But if one wants the greatest amount of honey a hive can produce he must be sure to place the first super—and all other supers for that matter—exactly at the right time because a hive that swarms rarely stores as much honey as one that does not swarm, all other factors being equal.

27. Q: *When would you place the second and third supers?*

A: The second super usually goes on our strong hive three weeks after placing the first one, and the third two to three weeks after the second. There is variation in this time interval, of course, due to varying weather patterns from year to year. After placing a super the head count of bees landing during the warmest part of the day usually drops from 110 bees per minute to 100 or even a few less than that. Then as the days pass the count gradually rises again until at the end of three weeks there are 110 or even 115 bees once more coming in for a landing. At this point, if the weather is at all favorable, we put on another super. After three supers are above the queen excluder we often see a great increase in the number of bees landing per minute and then we begin to add supers more often.

Counting bees is relatively simple especially if one has a partner to watch the clock and give one the start and stop signal. My father usually keeps track of the time and I do the counting. I often divide up the entrance area by using little white sticks laid at right angles to the entrance at intervals of about two inches. This divides the width of the entrance into little alleyways much like those in a bowling alley and for one minute I count the bees crawling hurriedly down one alley, and during succeeding minutes count the bees running down each of the other alleys. I add the various totals to find the grand total from the entire width of the hive for one minute. Such a bee count is amply accurate for all practical purposes.

28. *Q: Do you wait until fall to take off honey or do you take some off even during the honey flow?*

A: My father and I always take off a goodly part of our honey before fall. In most areas where we have lived the first frames and supers of honey have been filled, sealed, and ready to come off the hives three months after they were put on. I keep a running record on a calendar as to when we put on a super and from that can estimate when it should be ready to come off. But we try not to take any honey off the hives before each one has at least six or seven supers above the queen excluder for the reason that we want to encourage our bees to fill the empty supers that we add. 🐝 Bees must have the incentive of a feeling of well being and abundance before they will do their best work. But at the same time we must be sure that they have empty supers for living and storage space. When a beekeeper has acquired expertise in this phase of the work he is well on his way to becoming a master in the art of beekeeping.

29. *Q: What is a good bait for mice in a honey storage area?*

A: Bait your trap with a piece of walnut meat from an English walnut. In my honey storage room mice will spurn any offering of cheese but none can resist a nibble at a bit of walnut meat—so I catch mister mouse—and all of his relatives too.

30. *Q: Do you still use Mrs. Stewart's Liquid Bluing as a bee sting remedy?*

A: Yes. And use it liberally! Just a few dabs will not do the work. Saturate the sting area and keep it wet for at least five minutes, or even longer in some cases. We have many motorcyclists who pass through our county and sometimes one of them runs into a bee and gets stung. If they are within ten miles of my place they come roaring over to see Ormond and I apply a liberal amount of bluing. In five or ten minutes they wave goodbye and go on their way rejoicing.

Even more effective is Sting Kill from Diamond International Corporation. Diamond's Beekeepers Supplies catalog for 1977 reads: Sting Kill—relief from bee stings, mosquito bites, within one minute. Just apply Sting Kill until the germicide solution penetrates the skin. Packed ten swabs per box. Cat. No. A-407, one box Sting Kill swabs shipping wt. eight oz. Write the company for price and shipping charges. The company does not guarantee the product but I am told that it is effective and it is. I just now treated Dr. Charlie Scibetta for two bee stings near his eye and it worked like magic for him.

Juanita Polk of Powell, Tennessee recently wrote me a beautiful letter reminding me that honey is one of the best sting remedies. Indeed it is for many people. I have used honey but for so long now have been almost immune to bee stings that I had forgotten. But before applying any sting remedy we should be sure to properly remove the stinger. Always scrape the stinger off the flesh with a fingernail, knife, or even a scissors blade. *Never* pull out the stinger with a tweezers! That is the worst possible thing that one can do for it squeezes the poison sack forcing all of the venom out through the stinger and into the wound. Scraping off the stinger stops the flow of poison while the stinger is being pulled out.

31. *Q: Do you like a hive so constructed that the entrance opening can be adjusted in size?*

A: Yes indeed. I have noticed that some home craftsmen make their beehive entrances very large, as much as one inch or more in height and the full width of the hive and they leave them this way the year around. In certain areas I approve of such large entrances when the weather becomes very warm or the hive has built up in numbers to 100,000 bees or more. But until the weather does become hot or the hive exceptionally strong this large entrance should be kept restricted for the reason that in cool weather a cold wind blowing toward the entrance has a tendency to chill the brood chamber causing the queen to diminish her egg laying. In warm weather after the honey flow is past such a large entrance invites robber bees to attempt to invade the hive and they are too often successful as the guard bees have extreme difficulty patrolling such a large opening. Those beekeepers who already own hives with large openings should restrict them during the winter and early spring months. This is easily done by placing short blocks of wood on each side of the entrance. The remaining opening in the middle is then two or three inches long by the full one inch or more in height. But in my experience a far better way is to make entrance cleats that not only restrict the large entrance at each end but also make it lower in the middle. I take a sixteen inch long piece of one-by-three board and cut out on one side of it a notch one-half inch high by about eight inches long. I place this board or cleat as we call it in front of the hive entrance and the notch serves as a new and much smaller entrance. Then if need be, due to cold weather, I restrict this entrance even more with little blocks of wood at each end of the opening. I never have to nail or secure the entrance cleats or little blocks of wood to keep them in place because the bees quickly glue them in place with propolis.

The reason I like this latter type opening is that it gives the same overall square inches of opening but makes it far easier for the bees to ventilate their hive due to the air intake and air outgo being so much farther removed from each other. Also mice sometimes like to enter a beehive in winter and a long low entrance is much more difficult to enter than a short high one. My father and I make various size notches in our entrance cleats, and as winter comes on we place more restrictive cleats in front of those already in place until we have what we call "igloo entrances." These work well for us as they keep cold air in the brood chamber to a minimum and our bees come through the winter healthy and strong. Illustrations of entrance cleats may be found in our book *The Art & Adventure of Beekeeping*.

Beginning beekeepers sometimes bring me new factory manufactured beehives that have very low full width entrances. If I see that the maximum opening is only one-quarter inch or three-eighths inch in height I strongly suggest to them that we take the hive apart and add a three-eighths or one-half inch additional ripping all around three sides of the bottom board to raise the entrance opening to at least a full three-quarter inch height before bees are hived into the box. Then later on in the season when the bees build up strongly the owner can remove the restrictive entrance cleats and really have an adequate opening for the bees to use. In our experience a low entrance always means a poor honey yield as the bees are working at too much of a disadvantage in trying to ventilate their hive. It takes a vast flow of air past the open ends of the nectar filled cells to convert nectar into honey.

32. *Q: Do any of your experiments fail?*
    A: Unfortunately yes. And when they do fail it always costs us money as that particular hive is never able to produce as much honey as it would have if we had not meddled—or if we had been successful. Here is an example.
    Since there is considerable extra space still available in the standard hive body after the usual ten brood frames are in position, why not shave down the shoulders on the upper ends of the frame ends leaving them a total width of one and one-quarter inches instead of the usual one and three-eighths inches, shove them close together and add an eleventh frame to the brood chamber? Experiments had shown that bees would efficiently raise brood in the slightly narrower spaces available. So we reasoned that an eleventh frame would increase the bees' brood rearing area by almost one-tenth thus causing them to raise more brood to produce more honey. The theory was excellent—the actual results disastrous.
    "Well why didn't it work? It still sounds like a good idea to me," one man observed as I told him of our experiments.
    "Because to make space to add the eleventh frame we had to shave down all of the shoulders on the end pieces of all of the frames. This brought the top bar on each of the outside frames very close to the wall of the brood chamber on each side of the hive, leaving very little space for the bees to use to ventilate the hive as it was difficult for them to force moisture-laden air from the upper supers down past the queen excluder and through

those narrow side spaces. The result was that the bees gave up in despair and completely sealed off with a continuous band of propolis those useless too-narrow spaces between the top bars and walls."

"What happened then?" he asked.

"The bees became overcrowded in the partially sealed off brood chamber and threatened to build swarm cells."

"What did you do about that?"

"Well, we had been watching the bees for exactly one month and had seen that as far as drawing out comb and storing honey in the supers was concerned, they weren't prospering. So we took the hive apart, removed the eleventh frame, and respaced the ten remaining frames to their correct positions. We made the change just in the nick of time too for the bees didn't swarm. But neither did they make us as much honey as they could have because we wasted most of a month at the height of the honey flow. But we still harvested 120 pounds of honey and that is not too bad for a very dry year."

33. *Q: Why in the world do you use so many supers on your hives?*

A: Because we need that many. The reason, in a nutshell, is that during the honey flow our bees can collect in the daytime far more nectar than they can convert to honey at night. And since bees have no pockets it is we the beekeepers who must give them additional storage space as needed. It has been my observation that most of the beekeepers I know could obtain more honey if they owned and used more supers on their hives during the main honey flow. Too many beekeepers wait until the honey flow is half past and the bees have begun to swarm before they get around to putting on even one or two supers.

We need many supers because our nectar flow often begins January 1 in our Santa Cruz coastal area and lasts until July 1. During this period the field bees bring in nectar faster in the daytime than the hive bees can convert into honey during the night even though they sometimes run a night shift all night long. When we take an exploratory peek into a hive we usually find that many cells are almost filled to capacity with either nectar or honey but few if any are sealed. Then the only logical course of action is to add another super so that the bees have readily available empty cells already drawn in which to store the nectar the field bees bring home the next day. But if we have no drawn combs we add a super of frames with starter sheets only, placing this super down directly over the queen excluder, or at most one or possibly two supers above it. 🐝 A week or two after placing such a super it often pays to check the hive again and if we find four or five of the center super frames drawn we move them into an outside position near the walls of the super and move the empty undrawn outer frames into the center of the super. If we have some drawn frames we often place two drawn frames on each side of the super and six frames with starter sheets in the center. Bees like this arrangement too.

The year 1976 was dry and quite cool until well into the month of May. As a result the hive bees were totally unable to cope with the great quantity of nectar brought in each day. So we added supers and more supers until our largest and strongest hives had a total of eleven medium depth supers above the queen excluder. I had to work those hives from two stepladders placed about eight feet apart with a two-inch by twelve-inch plank nine feet long placed between them at a height of four feet. Even then I sometimes had to use a fourteen-inch high little bench on top of the plank so that after I had removed the cover I could look down into the hive sufficiently to see how the bees were working on the top combs. It pays to be somewhat of an acrobat when working such tall hives. If any of you can think of an easy way to work hives that are more than eight feet high I wish you would write to me and tell me your idea. My address is Ormond Aebi, 710 17th Avenue, Santa Cruz, California 95062.

34. *Q: Are there honey plants in California that are injurious to bees?*
   A: Yes. But only two that I know of—buckeye and dogwood. I have never had bees in a dogwood area so cannot dis-

cuss that phase of the problem. But the strip of buckeye trees that stretches in a narrow belt in the low foothills almost all the way around both the Sacramento and San Joaquin Valleys of California has always interested me as pollen from the buckeye blossoms is poisonous to the brood of honeybees. Yet for many years my father and I have observed bees working on these deadly trees. At first we thought that possibly the bees had always been brought into such an area by an uninformed beekeeper who in due time would sustain a severe loss. But always the next year we would still see bees busily working on the same buckeye trees. Whether they were the same bees as those we had seen the year before we could not positively ascertain, but they certainly looked like them. Then for a period of years we ourselves had about twenty colonies in a semi-buckeye area, and we sustained no noticeable loss. If buckeye were as poisonous as both scientists and practical beekeepers had asserted that it was, why had we seen bees working year after year on the trees in other areas and our bees did not seem to be affected either?

For a long time we attributed our lack of loss as being due to the fact that at the time of the buckeye bloom there was always also something else attractive to bees in bloom in the same area. And I am still not sure but that this is an important factor in keeping bees healthy in such a locality. But there also seemed to be another factor involved that we could not pinpoint until recently when a former bee inspector from a nearby county told my friend Ken some very interesting facts about bees and buckeye trees. He said that it had been his observation through the years that if one moved beehives into a buckeye area sometime in the winter and left them there in the one location they would survive the buckeye blooming season with very little if any loss, provided they were not moved at any time but were allowed to remain in the same location year after year. Then the bees seemed to be able to acquire the same ability to utilize buckeye bloom without ill effect just as wild bees living in a bee tree. But if hives are moved into a buckeye area weeks or even a month or two before the buckeye begins to bloom there will be great loss. Again, if the hives are moved out during, or even sometime after the buckeye season is past, he said there would be severe loss due to poisoning. I had never before heard this phenomenon explained as well. It surely agrees with what my father and I have also observed. If this proves to be true in all instances then a great deal of good bee pasturage may become available for more beekeepers since the honey is never poisonous to either the adult bees or mankind. Buckeye trees may still pose too much of a problem for the commercial beekeeper but for the backyard beekeeper—and many of them are building homes in such areas—it would be a real boon if they could keep enough hives to supply their family needs.

35. *Q: Why do you so often say that bees are like people?*
A: Because in so many ways they are. Let me cite just one instance. When we set out our large kettle of cappings to melt down in the sun we cover it with a square sheet of clear plastic both to help retain the sun's heat and also to keep out inquisitive bees that invariably buzz around any honey pot setting out in the sun due to the warm sweet vapors emitted by the warmed honey and melting wax. We have to be very careful that there is no hole or weak spot in the plastic, for if the bees can find even the tiniest hole or beginning of a hole they will work for hours to enlarge it so that they can get down to the melting cappings and honey. For this reason I always take periodic glances at my capping melter when it is in operation. The instant a bee enlarges a hole enough to squeeze through it is dead, of course, due to the heat. Nevertheless other bees immediately squeeze and squirm to get through the same hole, and if successful, they too, die from the heat. They literally work their hearts out to accomplish something that will mean their deaths.

Now let us compare the above bees with some people. A few years ago I had a young acquaintance who worked hard and saved and skimped so that he could buy a high-powered motorcycle. He waited most impatiently for the day to come when he would be sixteen and one-half years of age and could take his driver's examination. That day finally came. He took and passed the examination and roared away at high speed for Big Sur, California, about eighty miles away. Though part of the road is very curvy he rode down there in one hour flat—only to crash on a last curve and be instantly killed. Does that stop other boys from buying motorcycles and rushing to their deaths? No way! Hardly a week goes by but the newspaper carries the account of another cyclist who has been killed or injured. Do you wonder that I say bees are much like people?

36. Q: *Have you ever seen a queen's mating flight?*

A: Yes. Drones never mate with a queen in the hive but always out in the open air on a sunny day about fifty feet up in the air. They fly at a moderate speed, but even so, it takes a good pair of farsighted eyes to follow their flight. In my experience the queen has been followed by three or four drones who fly close behind her and one of them is the successful suitor. As soon as they are mated the drone dies. With hundreds of drones in the area one might think that they would fly after the queen like the tail on a comet. How the drones decide which ones shall be in the mating flight I do not know.

37. Q: *How do you control wax worms in the drawn combs you store over winter?*

A: As hobby beekeepers we stack supers with drawn combs in a cool place after we remove them from the hive for the last time and then about November 1 we take them out of storage and look at each frame. If we see spider-like webs running across the face of any comb we remove that frame and place it in our freezer over night or in our refrigerator for three days. The cold will kill any worm or egg. Large beekeepers stack their supers and apply a special gas to kill the wax moths, worms, or eggs.

38. *Q: How is it that you can work so easily with your bees and others find it so difficult?*

A: Because basically I am working with my *friends* the bees, whereas many people seem to have the attitude that they are trying to work with their enemies and have to fight the bees to get honey.

39. *Q: How long have you done business with Diamond International Corporation?*

A: For the past thirty years, or almost all the time we have lived here in California. Diamond published its first catalog in 1916, the year I was born, and has been in business ever since distributing supplies throughout the United States. They stock and sell smokers, veils, extractors and almost everything else that one can think of in the line of bee supplies including new five gallon square metal cans, beautiful five gallon white plastic round containers with air tight seals good for storage, squeeze bars, window cartons, round and oval freezer jars, bee books, all kinds of wooden bee products such as hives—and a great deal more. They give tours of their facility when arranged ahead of time and like to show their products, and how they make them, to as many as possible. I have been through their plant and it is really something to see. I have not seen their latest display area but am told that it is like a supermarket of bee supplies, one of the only ones on the West coast today. Their whole business is built on mutual faith and trust between themselves and their customers—a truly winning combination for all concerned. For a price list contact their Apiary Department, P. O. Box 1070, Chico, California 95927.

40. *Q: Who all keeps bees?*

A: Men, women, and children in every walk of life keep bees. On June 6, 1977 I helped a Catholic priest, beloved Father Markey, with his two hives of bees, and during the same month a year earlier I discussed beekeeping with Pastor James Smith, a talented and much loved minister of my own Advent Christian denomination who is also a beekeeper. And between those two dates I have received much beautiful mail in appreciation for our book *The Art & Adventure of Beekeeping*. One of the first to write was Dr. William Abler, Humanities, Illinois Institute of Technology, Chicago, Illinois 60616, who is himself the author of a remarkably useful and easy to read book entitled *Shop Tactics* (Running Press, New York) a treatise on the proper use and handling of all kinds of home workshop tools including those for working with plastics, glass, and silver soldering. Since we will never live long enough to discover the answers to all of our problems by trial and error, I firmly believe in surrounding myself with authentic information that has been recorded from the experiences of others. Therefore I regard *Shop Tactics* as a necessity for all home craftsmen.

Dr. Abler had the astonishing goodness to read my first bee book and jot down all of the typographical errors he found in it so that I could have them corrected in subsequent printings. I marveled that he knew so much about bees until he wrote that he too is a beekeeper and also does much microscopic observation and experimentation with bees. Another friend, Dave Dobbins, and Tom Waters, his neighbor across the street, work their hives together. This makes beekeeping more interesting and rewarding for both families.

Recently it was my great joy to introduce my sweet bees to Bill Creecy, beloved pastor of my own local church. He had been quite hesitant about stopping to see my father and me during daylight hours when our bees are flying in great numbers. But a few days ago on a warm sunny afternoon he did stop to see us. I soon invited him to step out of the house to watch our bees at work.

"Aren't you afraid?" my father asked him.

"Not when I'm with Ormond," he answered.

I was tremendously pleased and admired his faith and courage. We can all go places and do things when accompanied by an experienced guide that would be dangerous or foolhardy to attempt without one. It is no sin to be afraid of bees. I respect the natural fear of bees that many people exhibit upon coming to my place and I provide protective gear for them when needed. As for myself, I have been called brave, but it does not take a brave man to accomplish that which he is not afraid to do. Pastor Creecy exhibited bravery, as do many others who come here. We watched the bees for some time standing within five feet of the nearest hive and not a single bee buzzed us in any way. They liked him too.

Others who keep bees are farmers, carpenters, craftsmen of all kinds, office workers, doctors, dentists, and even the superintendent of a deliquent boys' home. This man told me that my friend Dorothy Ames, of Ames Apiaries of Arroyo Grande, so strongly urged him to buy a copy of our bee book for his boys to read that he went right away to the nearest bookstore and bought a copy. To his surprise those delinquent boys were delighted with it and many of them read it completely through, probably the first book in their lives they had ever begun reading and read through to the end. They urged him to buy a couple of hives of bees which he did. His boys took to those bees with the delight of love at first sight. That they got stung once in awhile did not faze them in the least. He said they would do anything for him if he would then give them an hour off to go watch the bees.

So if *you* feel frustrated, disheartened, and restless—take heart—it may be that you too have a delightfully severe case of as yet undetected bee fever and you need to observe and partake of the unbounded God given joy of life exhibited by our industrious little friends the honeybees.

Appendix B

Equipment Needs For A Beginner's Outfit

1-Story, 10-frame Hive.

2 Medium Depth Supers. We make, like, and use special supers that are ¼ inch shallower than the standard 6 ⅝ inch deep medium depth super box.

1 Queen Excluder. We use the wooden rim excluder if available.

Nails (galvanized), ½ lb. 7D Box for nailing hive and super boxes.

Nails, ½ lb. 2D or 3D for nailing frames together.

10 Sheets Foundation. We use 9 vertical, hooked wires pre-embedded at the factory into pure beeswax foundation starter sheets for the brood chamber. Hooked end goes up under pressure plate. Use 8⅛ inch sheets for solid bottom rails, 8½ inch for split.

20 Sheets Starter Wax for medium depth supers, 5⅝ or 5¾ inch width. We use either vertically wired or unwired beeswax starter. Bees like fresh starter. Buy a new supply each season.

Bee Brush.

Folding Veil, if purchased. Homemade—piece of window screen 3 feet long × 7 inches wide, sew ends together, sew on a cotton cloth top, sew on the bottom a piece of cotton cloth 3 feet long × 16 inches wide that hangs down all around and tucks into jacket top.

Bee Smoker. Get the big one with 4 inch × 10 inch barrel.

Hive Tool. A homemade one may be made from a piece of flat strap iron 1/8 inch thick × 12 inches long × 1 inch wide—bend last 2 inches of one end at right angles, and hammer a fairly thin and sharp chisel edge at both ends.

Boardman Feeder. A homemade entrance feeder will do in most cases, the easiest to make is a plastic coffee can lid with little wooden slats that float on the honey in the feeder to keep the bees from getting mired down while they lick it up.

Hive Staples. Used to secure the hive bottom board to the hive body. If one nails the bottom to the hive body it is difficult later to detach it, whereas hive staples are easy to remove with a screwdriver, hammer, or whatever is handy.

Gloves. Canvas, soft leather, or household rubber gloves as I use. Bees can sting lightly through any of these but seldom do enough to really hurt one. Heavy stiff gloves are so awkward to use.

Hive Stand. We like a stand 16 inches high, 5 feet long, and 2 feet wide built of whatever strong lumber is available as hives are heavy. This stand accommodates two hives, one near each end, with room in between for setting down empty or filled supers as needed.

Appendix C

Swarm
Hiving
Kit

Fully assembled hive with all 10 brood frames complete with wired foundation sheets.

Bee veil (I always carry two in case the swarm is high and I need help).

Gloves (I like household rubber gloves).

Smoker (4" × 10" barrel), cubic foot box of smoker fuel (burlap), matches.

Hive tool, pliers, 20' of light gauge wire or twine.

Bee brush, dipper, two gallon cardboard carton.

Hammer, sharp handsaw, twig clippers, sharp knife.

2 pieces of heavy corrugated cardboard 2' × 2' in size.

1 piece of heavy corrugated cardboard 3' × 4' in size.

2 saw horses, 2 pieces of 1" × 6" board 4' long.

1 piece ⅛" plywood 2' × 3' in size.

Pry-bar (I have one custom made with an extra wide blade).

2 lengths of ¼" nylon rope 40' long.

2 6' stepladders. Longer ladder should be available.

Screened wooden hive closure and a few 5-penny box nails.

3 or more 1" × 2" strips of board 18" long.

Gallon of drinking water and a little food.

For going after swarms there is nothing better than a large older station wagon with a car top carrier. During most of the swarming season I leave my car loaded and ready to go almost at a moments notice. Thus equipped when I go for a swarm I usually come home with it. However, at the present time I rarely go out for a swarm but instead transfer the call to one of a number of younger people who want bees to build up their apiaries. These young people, particularly if they have not as yet assembled a complete swarm hiving kit, will do well to ask the person who notifies them of a swarm, the following pertinent questions:

How long have the bees been clustered in their present location?

How high is the swarm from the ground or other substantial footing such as a porch or deck?

How is it situated on the building, bush, or tree?

If the swarm is high is there a ladder available on the premises?

Can you give specific directions for finding the property where the swarm is located?

Can you give me exact directions for finding the swarm if no one will be present to point it out?

The answers to these questions will determine the minimum of equipment needed to effect a successful hiving.

Appendix D

Hive Construction

Our world record beehive was constructed in the same way that we build all of our beehives. The accompanying drawings show our simple method of making all of the parts, which may be easily assembled, and work exceptionally well for us in the production of a maximum amount of honey. Figure 1 shows our brood chamber built on the "half lap" design which is much easier for the home craftsman to cut than the usual factory manufactured "dovetail" corners. Also we are convinced that in our damp cool climate, using unpainted hives, we get just a bit more honey by using the half lap design.

An even simpler design is that of the box hive in which the end boards are cut 14¾ inches long and the side boards 20 inches long. The end boards are placed between the side boards and all nails are first driven through the side boards and then on into the end boards so that the overall dimension outside measure is 16¼ by 20 inches. The top and bottom hand rails are cut the same dimension as given in Figure 1. When these are secured in place with two six-penny nails on each end and one in the middle, the hive box is really quite well double nailed in each corner. Some commercial beekeepers in California use these box hives. A six foot length of board carefully measured out will build either our half lap type of box or the box type hive but it is not long enough to build a dovetailed type box. All of our rabbet cuts can be cut on a small bench saw with a 7¼ inch diameter saw blade. We use seven-penny galvanized box nails to secure the corners, a total of twelve nails being required for each corner which total includes the four nails used to secure the ends of the handholds.

Figure 1 also shows our special medium depth super construction. The only deviation from standard manufactured supers is that we make ours only 6 ⅜ inches deep instead of 6 ⅝ inches deep. We like the shallower depth because it allows us to use 5 ⅝ inch starter sheets of wax which are readily available from all suppliers. When the day warms up we then bend each wax sheet over one quarter inch all along one side so that it will fit securely under our pressure plate (also called wedge bar) and be permanently fastened in place, and at the same time the lower edge of the sheet fits nicely down between the split bottom rails.

Figure 2 shows all of the cuts necessary to make the frames. All of them may be cut on a bench saw, but it is faster and easier if one also has a four inch jointer planer to use in combination with the bench saw. Even the two notches at the lower end of each frame end may be cut with the bench saw by using a dado head. A serviceable dado head may be made from a worn down saw blade. We secure the blade firmly in a vise and then, with a monkey wrench, we bend a one-eighth portion of the circumference of the saw blade over one-eighth inch to one side. The next one-eighth part of the circumference is bent one-eighth inch in the opposite direction and so on around the saw until all eight bends are complete. We now use two such blades properly spaced thus enabling us to cut both notches at one pass of the saw. We use two-penny nails to assemble the frame parts.

WaveCloth

Box Type Super

Hive Construction

Figure 3 shows our simple method of hive cover construction. The advantage of this type is that we can slip it onto the hive from the side and at all times observe the bees. With this cover we rarely crush any bees for we can brush them aside or push them aside as we slip the cover onto and across the top of the hive. The end rails serve as a guide in correctly replacing the cover. We always use an auxiliary wooden or metal cover as a roof so we do not need to use a telescoping cover to turn rainwater.

Figure 4 gives the details of our bottom board. We always run the boards lengthwise of the hive to make it easier for the bees to keep their hive clean, and also in the case of a severe storm the rainwater will drain out by gravity. We keep the rear ends of our hives one inch higher than the entrance.

A most important feature of our world record beehive, and of all of our beehives kept for maximum honey production, is the extended landing board. The extended landing board is placed immediately in front of the entrance on an exact level with the entrance thus allowing the bees as much as eleven inches beyond the hive entrance to gather in greater numbers to efficiently ventilate the hive and draw damp moist air from the interior of the hive.

An extended landing board is easily affixed to a beehive. We place two one-by-four-by-thirty-six inch boards, one on each side, under the outer edges of the bottom cleats in a horizontal position running lengthwise of the hive, and let them project about eleven inches forward beyond the entrance. When these two boards are laid in place and the hive placed upon them the weight of the hive holds them securely in place without other attachment. The addition of two two-by-six-by-eighteen inch boards laid flat in front of the hive entrance completes the extended landing board. The entire hive should be built as nearly as possible in such a way that it will provide maximum efficiency for the bees. Remember that a beehive is a combination bee nursery, housing development, and honey factory.

Hive Plans

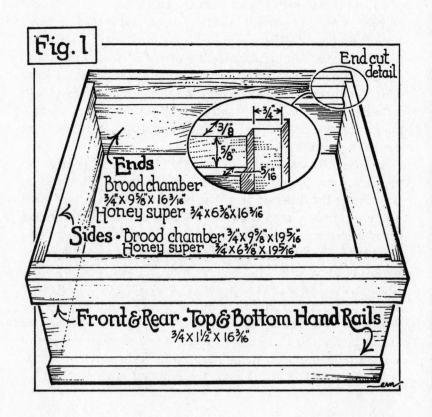

Fig. 1

End cut detail

Ends
Brood chamber
¾" × 9⅝" × 16³⁄₁₆"
Honey super ¾" × 6⅜" × 16³⁄₁₆"
Sides · Brood chamber ¾" × 9⅝" × 19⁵⁄₁₆"
Honey super ¾" × 6⅜" × 19⁵⁄₁₆"

Front & Rear · Top & Bottom Hand Rails
¾" × 1½" × 16³⁄₁₆"

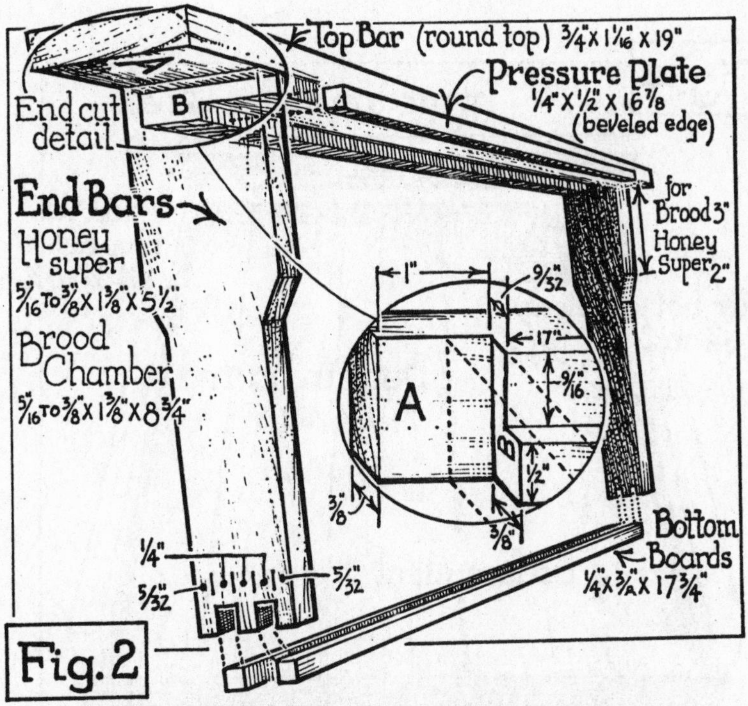
Top Bar (round top) 3/4"x 1 1/16" x 19"
Pressure Plate
1/4" x 1/2" x 16 7/8
(beveled edge)
End cut detail
End Bars
Honey super
5/16" to 3/8" x 1 3/8" x 5 1/2"
Brood Chamber
5/16" to 3/8" x 1 3/8" x 8 3/4"
for Brood 3"
Honey Super 2"
1"
9/32"
17"
9/16"
A
B
1/2"
3/8"
3/8"
Bottom Boards
1/4" x 3/4" x 17 3/4"
1/4"
5/32"
5/32"

Fig. 2

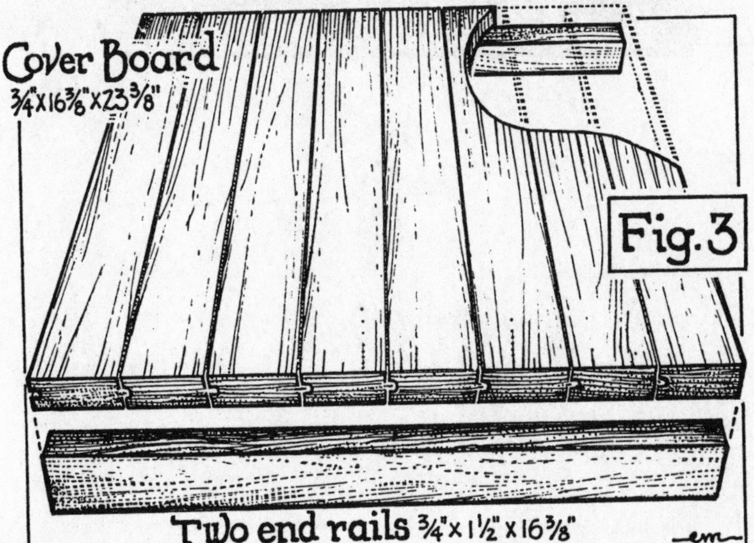
Cover Board
3/4 x 16 3/8" x 23 3/8"
Two end rails 3/4" x 1 1/2" x 16 3/8"
em

Fig. 3

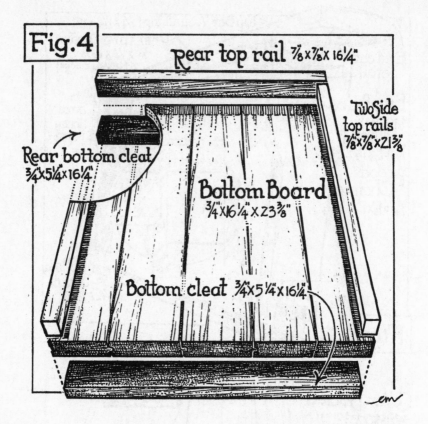

Fig. 4

Rear top rail ⅞ x ⅞ x 16¼"

Two Side top rails ⅞ x ⅞ x 21⅜"

Rear bottom cleat ¾ x 5¼ x 16¼

Bottom Board ¾" x 16¼" x 23⅜"

Bottom cleat ¾" x 5¼ x 16¼"

Hive Plans

The Honeybee's Hope

Ants! Ants in the entry—
  More ants on the stand!
Even ants on the landing—
  And all over the land!

Help! Help! Sister bees
  Cry the guards by the door
We're invaded by ants!
  They're all over the floor.

Now they're climbing the combs
  Cries a poor tired worker
Send guards to the queen
  They must not harm her!

She's our light and our life
  Oh what can we do?
The ants will have robbed us
  Before they are through.

Call out the hive bees
  Though sleepy or not
We'll fan with our wings
  And blow out the lot.

And fan they did do
  With their tails pointed high
And impeded the progress
  Of all the ants nigh.

But we never will win—
  Cried the queen in distress
We must quickly have help
  From our keeper, no less!

Send a call out for Ormond
  He's a good man and true
He has ears like a fox
  And will know what to do.

So they sent a clear call
  And as Ormond came round
He instantly dropped
  On one knee to the ground.

Aha, he said softly
  As he saw the invaders
We'll make short work here
  We'll poison you raiders

Take heart little bees
  There's help on the way
I'll dispatch these ants
  And save you today.

The bees heard their keeper
  They buzzed round in delight
Knowing all would be well
  Before the kiss of "Goodnight!"

Honeybee Serenade

Ormond Aebi   HONEYBEE SERENADE   LaVerne Faris

1. On a sun---ny spring day in the mon---th of May
2. O--------- lo----ve those bees 'each----- one----- is dear
3. ----------- Ho---ney sweet honey th---ey gath--er by day

My bees take wing and hum a---- way In---to
Soon they re-----turn with nec---tar clear Back to
Work and refine it the whole night away Love----

the az---ure blue----- sky------------- Grate- ful to God
their hive and en----ter in---------------- De-- posit their loads
those bees---- for the gold they store A gift of life

for the pow-----er to fly. Buz---in' by the doz-en
and go out-------- a---- gain. Thousands and------ thousands
and------ health--- and more. Buz--in' in the flowers

all day long Hum--min' in the sun a joy-ous song.
toil without fear Pre-cious lit--- tle insects so free and so dear.
all day long Bringin' in the honey a -- singin' a song.

/ 273

Index
To
Beekeeper's
Questions

Index

Note: The page numbers in bold-face refer to pages in *The Art & Adventure of Beekeeping*.